I0505873

Reign of the Chroms

by

Zoltan Bartok

Part One

"Right on time," said the young man when Lella entered his chamber. "Are you ready?"

"As ready as I can be," answered the woman. "What?" she said a moment later seeing the concerned look on Ravar's face.

"Sit down for a second," the man replied after a brief silence. "There's something we need to talk about."

"Well." The woman frowned as she put down her shoulder bag. "I hope nothing serious."

Ravar paused again. "Lella, are you sure you want to go through with this? Think it over one last time. Once we leave, there's no turning back. Even if we wanted to, the Shuttle could never catch up with the Arrow. Also, there's no guarantee we succeed finding what we're going to be looking for out there. The Shuttle is not very fast, it could take a long time to find a habitable planet."

"I have nothing to think about," Lella said resolutely. "The sooner we leave the better."

Ravar didn't answer right away. "There is a moral issue here as well. It's possible that the Arrow won't last much longer. The reactor could explode in a few days. When the readings begin to confirm that this is inevitable, there will be others, I'm sure, who would appreciate the chance to escape. Once we take the Shuttle, no one will have that chance. We'll have to live with the burden of that.... Are we going to be all right?"

"If you say yes, I say yes," Lella answered.

They looked at each others for a while.

Finally, almost simultaneously, both of them nodded.

"All right then." The man stood up first. "Let's go!"

"Wait!" Lella said with a broad smile. She tiptoed and gently kissed Ravar's lips.

"Is your plan as good as I think it is?" Lella asked while they were crawling through a long tunnel. She finished her question with a few coughs.

"You'd better save your breath," replied the man. "There isn't much oxygen in here."

"Is it?" Lella insisted.

"Well, we may have to improvise here and there."

"I see," acknowledged the woman. "I hope you'll know when."

When they arrived at the end of the tunnel, Ravar opened his toolbox.

"Here we go," he sighed after unscrewing a section of the floor. "Down there!" He looked through the opening. "That's our shuttle."

"Wow!" Lella gazed at the vehicle.

Ravar tied a rope to a metal bar and they both lowered themselves to the hangar.

The man punched in some codes on his palm-coder while Lella kept looking around nervously.

The shuttle door opened and they jumped inside.

Ravar immediately started the engine.

"Won't you shut the doors?" Lella asked anxiously.

"They'll close in a moment…. Computer! Accelerate!"

A faint vibration ran through the structure of the vessel.

"Engine in operation," reported the computer.

A red light started flickering and a warning signal sounded, then the computer spoke again: "Warning! Exiting the Arrow now may tear the Shuttle apart."

"Explain!" Ravar snapped right back. He was losing his cool.

"The Arrow is at its possible maximum speed. The Shuttle is not meant to be launched at such high speed."

The man stared at the woman.

"Don't ask me!" Lella said in a hurry. "Do what you'd do if you were alone."

"Computer, open space door!"

"Door is opening."

Overpowering the voice of the computer, a harsh alarm sounded. "Warning, warning!" roared the alarm. "Exiting the Arrow may destroy the Shuttle."

Ravar hesitated.

The next moment, very loud sirens went off in the hangar. As they looked out their window, Ravar and Lella could see members of the security team storming into the hangar.

"Computer," Ravar yelled as loudly as he could. "Proceed with launch!"

The Shuttle gently lifted off its pad. As soon as it started moving forward, Lella and Ravar were pushed back in their seats.

"Full speed ahead," Ravar shouted.

As they cleared the exit, the Shuttle began to shake violently. Ravar's eyes were on the side window. He saw the Arrow disappearing in the distance as their vehicle was rapidly decelerating.

Lella was thrown from her seat. She hit her head and lost consciousness.

Ravar was able to hold on. He checked the control panel and saw that everything was returning to normal. He sighed and then he instructed the computer to take control. "Scan for life signs!" he added. "And maintain maximum speed!"

"Command accepted."

Ravar kneeled next to the woman. "Lella, can you hear me?"

There was no answer. The woman was motionless. Her forehead looked badly bruised and one of her shoulders appeared out of place.

"Hold on," said Ravar. He reached into his bag and pulled out a small box. He activated the device and then he held it against Lella's forehead.

4

It did not take too long before the woman opened her eyes.

"How do you feel?"

"Like being lost in space," Lella replied. "Dizzy... definitely dizzy."

"Your head looks fine." Ravar moved his device to the woman's shoulder. "Well, this shall kill the pain. Now, I'm going to make a slight adjustment here.... Just like that.... Looks better already. Don't move yet! Let's wait until the numbness is gone. If you feel no more pain, you're as good as new."

Lella smiled. "I've never had a better doctor than you."

"Wow! This was easier than I'd thought." Ravar sighed with relief after the woman stood up.

Back in her chair, Lella glanced out the window. "So then, what's our situation? I see the Shuttle is still in one piece."

"Lella, we're free!"

"Could you be somewhat more specific? For example, I'd be delighted to hear what's going to happen next."

Ravar did not answer right away. "Lella, this is learn-as-we-go. I know how to pilot the Shuttle. I have a few other skills as well. What I don't know is how to predict the future. Please, be patient. I'll do my best to get us somewhere."

The woman slowly turned her head away. She was once again looking out into the vast darkness dotted with millions of tiny spots of light.

Ravar got himself busy studying the control panels.

"I'll try to be as much help as I can," Lella said later.

A beeping signal sounded as a screen turned on. First, the screen was blue and then gradually, mosaic by mosaic, a woman's face appeared.

"The Captain," Lella said in great surprise.

The picture kept breaking up, some or all of the face disintegrated for short periods of time.

"She's talking to us," Ravar noted somberly.

"Why can't we hear a word?"

"They're too far away already.... Besides, their transmission is not perfect. I'm wondering how it is that they are able to send this picture at all. Yesterday, that wouldn't have been possible."

"Do you think they're making progress bringing the Arrow back on-line?"

"It's possible."

"So, what if they can correct all the problems? Can they find us? Would they take us home? Would they.... punish us?"

"Lella, I refuse to speculate. Right now, we're facing an entirely different reality.... In fact, look here! Look at this screen! It's showing a huge cloud of cosmic dust right ahead of us.... Computer, reduce speed!"

"Speed adjusted by minus ten percent."

"Activate remote scanner!"

"Remote scanner activated."

"Define nature of matter ahead of us as far as possible!"

"Analyzing..."

It was quiet again for a while. Only the faint buzzing of the Shuttle remained.

Lella kept switching her glance back and forth between the panels and the screen. Eventually, the Captain's face faded out completely, and the screen returned to its light purple stand-by color.

It was the computer's voice again: "The matter ahead is a galaxy. Its size is insignificant by Omega standard."

"Any life signs?" inquired Lella.

"Unable to determine from current position."

"Adjust course to penetrate mid-region," Ravar said. "Maintain speed as long as it is safe.... Report any signs of life."

"Commands accepted."

"What's our direction in relation to the Arrow?" Lella asked.

"Unable to determine," replied the computer. "Arrow is no longer within scanning range.... Last recorded direction.... 164 degrees off Arrow's course."

"Out of scanning range," commented Ravar. "No wonder. We're going the other way. Well, almost the other way."

"Couldn't we just find our way home in the Shuttle?" Lella asked.

Ravar released a brief laugh. "Yes, in about three of our lifetimes, combined. Besides, the Shuttle would never make it through the barrier outside our segment."

"Our food rations wouldn't last that long, either," added Lella.

"That's right."

"Hey, aren't you getting hungry? I could eat something."

"Here's your first chance to give the mixer a try. It's right there behind you. I don't want anything yet."

"And what can we have here for entertainment?" Lella asked after she finished her meal.

"Anything, as long as it doesn't require us to leave our seats... We can listen to music, watch plays or sports events. Ask the computer to read a book for you... or teach you about ancient history."

The computer interrupted him.

"The only life form detectable within the mid-section of this galaxy is directly ahead of the Shuttle. Distance is approximately half a light year from here."

"Boy!" Lella shouted joyously. "That's great news."

"Terrific!" Ravar clenched his fist and punched the air. "Computer, how long will it take before we reach the planet?"

"At current speed, we will arrive at the outskirts of the solar system in half a day."

"Continue on course!"

"Acknowledged..."

"Lella, before you know, you'll have solid ground under your feet."

"I'm looking forward to that.... However, now that I've fed myself, I feel like taking a nap."

After the woman entered her sleeping chamber, Ravar selected a documentary nature presentation about the deep jungles of their home planet.

"On screen!" he instructed the computer.

Once the show began, he reclined his seat and made himself comfortable. Later, he turned off the sound. He just couldn't keep his eyes open any longer.

When Ravar woke up and looked out the window, he saw a bright star in the dark universe.

"Computer, report current position!"

"The Shuttle is orbiting the solar system you set course for. The autopilot was unable to enter the system without specific instructions. There is too much debris flying around the central star."

"What are the parameters of the planet with life?"

"Code designated for that planet is Y2122. It is a small planet, about one tenth the size of Omega BX. It is relatively close to its star. Consequently, it circles the star in a very short time. Its rotation around its axis is approximately one hundred times faster than that of Omega BX. The star, by the way, is also very small compared to the Omega star."

"Reduce speed to fifty percent."

"Speed is already at fifty percent."

"Of course, of course," mumbled Ravar. "Computer, set safest speed and take the Shuttle in. Set course for Y2122."

"Command accepted."

By the time Lella emerged from her sleeping chamber, the Shuttle had arrived at its destination. During the time she spent in the bathroom, Ravar completed all necessary maneuvers.

"Good morning," she greeted Ravar while she was ordering a cup of tea from the processor.

"Good morning to you, too, Lella.... and greetings from Y2122."

"What is Y2122?"

"Your planet, my dear, come and see."

"Oh, really, are we home?" Lella forgot about her tea and dashed to the window. "It's beautiful! What a nice, vivid blue planet! It looks better than Omega BX, don't you think?"

"I haven't thought about that yet."

"Oh, I like it a lot! But why are we so far away?"

"It's a very small planet, Lella. That's why it's only the size of a coin from out here. You see, if we were orbiting Omega BX, this would be a safe distance to do it. Obviously, the gravitational pull by Y2122 is almost insignificant in comparison. We'll take up a much closer position as soon as I'm able to determine what we're dealing with. However, right now, you'll have to excuse me. I'm in need of a shower. I'm hungry, and so on. See you later."

Lella looked at the planet. "I'd rather be down there soon," she said with a sigh.

"We'll work on that."

While Ravar was taking care of himself, Lella updated her diary.

"Let's start our investigation," Ravar said when he later returned to his seat. "Computer, start data extractors and scan surface. We need to know what kind of life forms we have on this planet. Provide all information as they become available."

"Scanning in progress," reported the computer after a short break. "Y2122 is populated by a large number of species. Bio-electronic readings suggest one species with high level of intelligence. This species number in the thousands. Their shapes resemble that of the inhabitants of Omega BX. One significant difference. They are about two thirds smaller."

"Does this mean that I'm three times taller and three times heavier?" interrupted Ravar.

"Yes, approximately."

"Are they a homogenous sort? Let's see the Chromosomes!"

"Analyzing.... There are four distinct groups. Names designated are Type A, Type B, Type C, and Type D. Type A and B live in the same area of a small island northwest of the big land mass."

"How many continents are there?

"There are three continents counting the above mentioned island as one. All three are on the side of the planet we are currently facing. The other side of the planet is completely covered by water."

"How about Type C and D?"

"Type C lives in the north, while Type D in the south of the largest continent. This continent narrows at the equator. That area appears to be a jungle with numerous volcanoes and other high mountains. The third continent is to the southeast. Its size is hardly larger than that of the one in the northwest. A very few of the Type D creatures live here."

"Type A is the oldest of the four groups," continued the computer. "The other three groups must have appeared on the planet at a later time. Members of the Type A group possess an extra Chromosome which can not be found in members of the other three groups."

"Is Type A more intelligent?"

"It's impossible to determine at this point."

Lella curiously watched on. When the computer paused, she stepped back to the processor to fetch her tea.

"You know all this," she said after taking a sip of her tea. "I'm amazed how well versed you are! Just like watching the Captain in action."

"I'm the captain," Ravar said with a smile.

"I meant the Captain back on the Arrow. Of course, you're the captain here. No one disputes that. I'm just wondering whether you'd have ever left the Arrow if you hadn't been demoted. After all, you were a very capable strategic officer. Some of the older commanders were jealous of your abilities. They were the ones who immediately accused you of subordination when you dared to challenge the Captain's orders on that critical day."

"Lella, thanks for the compliment. As for what happened there, let's forget it! I'm perfectly happy here with you. I know you well enough, I think. I know you're a rebel just like me. Even if the Arrow hadn't broken down, we should've come up with this solution for ourselves. We'd never be content in our rigid society. I'd thought that by leaving on a space mission, freedom could be achieved. Oh, no! The rules and regulations, the Protocol, hung over us even a million light years from home. I know you're homesick. So am I. You know what I think? I think we're both homesick for a real home. A home where we can be ourselves. I must admit I would've never left if I had to do it alone. It's because of you, Lella." Ravar paused. He stepped closer to the woman and reached for her hand. Lella reached out at the same time. "I only did it because of you, Lella. I love you."

"Oh, Ravar..." The woman moved tightly close and put her arms around him. "I've been waiting for this moment for a long time. You know that, don't you?"

"Lella, I'd go to the end of the Universe with you. I'm glad I finally have the occasion to tell you this."

"I love you, too, Ravar. I love you very much." She tiptoed and kissed him. "I would've followed you anywhere."

"We'll create our own world, Lella, you and I. We'll start a new civilization that'll be free. No Protocol!"

"No Protocol!" echoed Lella.

The computer interrupted them. "Readings from the surface indicate adequate supply of oxygen. Some of the vegetation is composed of the same chemical elements as the edible plants on Omega BX. However, they contain a wider variety of amino acids as well as fatty acids."

"What about weather conditions?" Ravar asked.

"The range of temperatures is much narrower than on Omega BX. This range falls well within our comfort zone."

"Wow!" shouted Lella happily as she let go of Ravar. "We couldn't possibly ask for anything more. This is a perfect place for us."

"It appears so," commented Ravar.

"What's wrong?" Lella asked, seeing the wrinkles on the man's forehead.

"It's just occurred to me, Lella. We do have a problem."

"A problem. What problem?"

"We won't fit in. Just think about it. We're three times as tall as they are. I mean the inhabitants of this planet. They're midgets compared to us. This alone is a big enough problem. Then, we have no idea of their behavior. They might be wild or extremely primitive."

"We could be their Gods," Lella said.

"Yeah, perhaps for a while. Sooner or later, some of them would want to get rid of us. We're grossly outnumbered, we can't ignore that fact."

"Are you telling me that we have to move on and find another planet?"

"No, I'm not saying that. In fact, scanners couldn't detect any other life signs in this third of the galaxy. This most likely means no livable planet, either. Unfortunately, we can't risk a trip to the extremities of the galaxy."

"Why?"

"I guess you should know. The self-diagnostics indicate raptures in the outer shell of the Shuttle. We must've sustained this damage at the time we exited the Arrow. Luckily, we haven't encountered any storms on our way to this place. Chances are we would if we moved on. The Shuttle wouldn't survive even a weak disturbance."

"What's the solution then?" Lella paused. "Should we just irradiate these creatures off their planet? We could do that, couldn't we?"

"That would be an easy task. Our gamma torpedo would do the job with a single blast."

"So? We were thinking about starting our own civilization anyway."

Ravar was deep in his thoughts.

"We leave Omega BX because we couldn't adapt," continued Lella. "Wouldn't it be even more difficult to adapt to primitive creatures? Besides, let's not forget our physical differences, either, as you also pointed that out."

"Lella," Ravar responded quietly. "Are you really suggesting that we destroy an entire civilization?"

Lella bit her lower lip while shaking her head.

"Not a good idea, is it?" She was sort of saying this to herself. "Besides, the gamma torpedo would destroy all life, including the vegetation, right?"

"That's right, Lella. Eventually, the vegetation would die, too. Without vegetation, there would be no oxygen."

"It's not a solution then, obviously. Do you have any ideas?"

Ravar nodded. "I think I do."

"Really. Can I hear them?"

"Give me just some time. Let me analyze these data." He moved over to a side panel and began to run some computations.

Lella stepped to the processor. "In the meantime, I'll prepare something for breakfast. Can I get you something?"

"You sure can. I'll have the same as you."

"So, what did you come up with?" Lella asked curiously when Ravar joined her.

They both turned their seats inward and used the top of the divider in the middle to support their trays.

"We'll make them grow."

Lella raised her eyebrows. "We'll do what?"

"We'll make them grow," Ravar repeated. "I don't mean physically. That we can't do. However, we can make them evolve."

Lella seemed puzzled.

"They would evolve anyway," continued Ravar. "We'll just speed up the process. Listen to this! The readings show that they live in an entirely different time continuum. This makes perfect sense given the quick rotation of their planet and the short time it takes for them to go around their sun. A generation probably comes and goes just while we're having this conversation."

"Are you kidding?" Lella's mouth remained open. "Now, wouldn't that make our life down there even more impossible?

Just think about it! They'd grow old while we told them what we had for breakfast."

"Not exactly. Once we leave our universe and join theirs, time passes the same way for us as for them."

"Ravar, stop for a minute! I don't want my life reduced to the length of a piano concert!"

"Don't worry, that wouldn't happen. Once we cross the time barrier, we'll perceive the passing of time differently. A lifetime for those creatures down there appears probably as long as our lifetime appears to us on Omega BX or in here. Be assured, we wouldn't die in an instant, and we wouldn't live forever, either."

"I'm not sure I can relate to everything you're telling me but I'll take your word."

"Good! So, where were we? We were talking about their evolution. Just a few days for us up here will see a good number of generations come and go on the planet. This in itself is no guarantee of any significant evolution in their way of life or in their mentality. Currently, modern technology is unknown to them. This could remain unchanged until the end of time."

"I see!" Lella said. "You want to inspire them to invent a new world for themselves."

"You got it!"

"It's not a bad idea, but how can you achieve this?"

"We'll use our bioblaster. Thank the Universe for the Type A group and their extra Chromosome. This group is capable of receiving the impulses our bioblaster can generate. We'll enhance their minds. They'll guide the entire population on the road to an eventually highly civilized society. Remember! A thousand years for them is only a day for us up here. How many thousands of years will they need until they reach the right level of understanding? Does it really matter? A few days here or there won't make a big difference for us. Sooner or later, they'll be able to tolerate us. In fact, they'll be glad to see us living with them. Why? It's because they'll be as intelligent as we are."

"I guess you won't have much time for me during these next few days then. You'll be quite busy playing the role of a creator. It'll be a great accomplishment if..."

"No ifs, Lella! I know it'll work."

"I'll let you get started. I'm here if you need me. I'll be reading some books on intergalactical relationships."

"Right on!" Ravar stood up and gently pulled the woman out of her seat. "Let me just have another one of your heavenly kisses and then I'll move us closer. We'll orbit the planet once to check out the other side. I'll fix our position as if we were another one of the planets orbiting their sun. The inhabitants will perceive us as a new star in the sky. We'll be visible to them only for a short time during the night, if they notice us at all."

Lella pressed her lips against Ravar's. They kissed for a long time. Finally, Ravar freed himself from Lella's arms. "A generation has just passed down there," he said. "And I've just realized something!"

"Again?"

"Lella, I think you'll like this one." Ravar sounded excited. "I am glad it occurred to me. Look here! Inside this computer, we've got a genie." He pressed a few keys on the panel. "Computer, activate the holorobot!"

"Specify gender."

"Male!" Lella gave the order in a big hurry looking at Ravar apologetically. "Sorry," she said, "I wouldn't be able to tolerate another female in this tight place. I would surely be jealous."

"We needed it to be a male, anyway. That's what I was going to request. Thanks for helping out."

Out of nowhere, a man appeared. He was standing in the passageway, dressed in the same uniform they were wearing, looking somewhat disoriented. He looked almost like Ravar, about the same age as well.

Lella was amazed by the sight. "You call this a robot? He looks real to me."

The hologram looked around. He tapped his own body at various places, then, with a jovial smile on his face, he turned to Lella. "Hello," he said in a pleasantly deep, masculine voice. "I'm Wrix." He extended his hand.

"How weird," Lella said, "he's standing right here, but his voice is coming from that computer over there."

"Wait!" Ravar called out. "He's going down to the planet. He can't be as big as we are. Computer! Reduce size of holorobot by sixty-five percent!"

Wrix shrank in an instant.

"Did you really have to do that?" Lella asked with some disappointment in her voice. "Now I have to lean down to shake his hand."

"I'm Wrix," repeated the robot as if nothing had happened.

"I'm Lella... and this is Ravar."

"Lella.... Ravar," the robot said after shaking hands with both. "I'm not familiar with the two of you."

"Of course, not," Ravar replied. "We've never met. You've been stored in a computer ever since the scientists created you."

"Well, I'm very glad to be freed, finally, thanks to you."

"Wrix, we'll need your help."

"I'm at your service."

"We'll beam you down to the surface of planet Y2122.... I suppose you have no information stored in your circuitry regarding Y2122."

"That is a correct assumption," replied Wrix.

"I'll program all relevant information into your host computer," continued Ravar. "You'll receive additional instructions as needed. Now, I'll have to ask you to deactivate yourself as a solid projection and then take over control of your voice."

Wrix disappeared.

"How is that?" Wrix asked. His voice was no longer coming from the host computer.

"Perfect. You're not to appear solid again until beamed back from the surface unless the circumstances require that you do. Make sure your holotransmitter never gets turned off. That way we'll be able to see on this screen everything you'll see down there. Voice transmission is needed as well. It'll all be recorded and stored for us to analyze any episodes. Are we ready?"

"Ready," replied Wrix.

Ravar pressed a button. Soon after that, looking at an amplifier, he noted: "All right, he's arrived."

"Great!" Lella said. "Now, can we finish that kiss?"

Ravar hugged her and held her tight.

"Look! Up there!" the young man shouted to his father who was busy trying to keep a wild beast away from the flock.

"Wow!" the older man reacted after he looked up. "It's a new star! It's moving across the sky!" As he said that, he completely forgot about the beast. He did not even hear the whining of his favorite lamb that soon became silent between the jaws of its predator.

"Pap, you see it?"

"I sure do, my son... I sure do... It's heading right for the top of Mount Ulvi... It's not one of those falling stars that zip through the sky from time to time... This one is moving slowly... and it's not burning up, either."

"Look Pap! It stopped."

"You're quite right, son. It's stopped, indeed."

"Now it looks just like the rest of the stars," the younger one added, "perhaps a tiny bit brighter."

"Son, this is a special night."

"It's going to disappear soon."

"I see, my son... I see... Why don't we give it a name before it leaves? Let's call it Brius."

"Brius... Do you think the folks in the village will like this name?"

"I hope so. After all, I'm their leader. As leader of the Chrom people, I have the right to name a new star, don't I?"

"The Chrom people?" the son asked startled. "I've never heard this name before."

His father pondered. "That's because we haven't used it for generations. I can assure you that it's our name... Chrom."

"How do you know?"

The father pondered again. "I... just know. It's just occurred to me."

"Interesting," said the son, his eyes still fixed on their new star. "It's going now. You see it, Pap? It's going behind the ridge now... It's gone."

"I see, son, I see. The day is breaking on the horizon, too. It's time for me to return to the village."

"I'll do fine, Pap. I'll sleep later, after the sun comes up. The flock will be safe then."

The father picked up his leather bag that stored his water, hung it at the end of a wood stick, put the stick over his shoulder, holding one end with his right hand, and started walking across the field, heading in the direction of the rising sun.

When he arrived at the edge of the village, some people were already going to the field to work on their parcels.

"Good morning, Dorug," they greeted him.

"Good morning... Good morning," he returned their greetings.

"Where is Zuppo?" asked a young woman who was carrying a bunch of long sticks on her shoulders. "Did he stay out with the herd?"

"That's where he stayed... He won't be coming back for a couple of days. Have a good day, Mirna."

Some people were pulling carts. Some had cows to pull them.

"Hey, Tiho!" he called out to an old man with a hump on his back. "Why don't you trade some of your wine for a cow?

You're getting too weak to pull that cart yourself. I see you're struggling."

"I'll think it over, Dorug. Thanks for the advice," the old man answered cheerfully.

"At least you should ask someone to push it for you," Dorug mumbled as the old man moved on.

"Hello Marsi," he greeted the white cat that always waited for him on top of the stone wall at the edge of the village.

The cat jumped off the wall and rubbed its fur against Dorug's worn leather boot, and then, as usual, disappeared through the holes of a wooden gate.

His wife was already anxiously waiting for him in front of their stone house. She was holding the gate with one hand and waving to Dorug with the other one.

"Good morning to you, Bria, my dear." He kissed her, wrapped his free arm around her waist, and opened the gate with a gentle kick. "I'm so glad to be home."

"I have fine porridge for you. Just finished cooking," the woman said after they entered the house.

The man settled at the bulky table he had cut out from the trunk of a huge tree. He used the rest of that tree to carve the seats.

"What's in this stone pan?"

His wife lifted the wooden cover. "It's fresh butter. I churned it last night. I thought you might like to have some on your porridge... And here's the milk." She put a clay jug on the table.

"We lost another lamb last night," the man said after he finished eating. He put down his wooden spoon and kissed his wife again. "The beast took it... But hear this, Bria! We saw a new star in the sky!"

"A new star... Is it still there?"

"It'll be... before the sun comes up tomorrow, I hope... I named it Brius."

The woman smiled affectionately as she moved closer to her husband.

"I like that name," she said.

Someone was knocking on the thick, heavy, wooden door. They heard a man shouting Dorug's name over and over again.

"Come quickly!" the man said when Dorug opened the door. "Your daughter is here."

The woman put her palms together. "My God!" she cried out loud. "Zea, my daughter... Where is she?"

"There." The man pointed at the gate. "Right there, outside."

Dorug and Bria ran as fast as they could.

When they opened the gate, they saw Zea lying there on the ground, bleeding from wounds on her legs, on her arms, even on her face. The rags she wore hardly covered her from her bulging breasts down to her thighs.

"Zea," her mother cried out.

"My daughter," Dorug leant down and took the girl in his strong arms. He hurried back into the house where he placed Zea on one of the bunks. "Water, she needs water."

Onlookers gathered outside peeking in through the open door.

Bria rushed back with a jug and a towel. She filled a cup with water and put it to her daughter's mouth.

The girl, with eyes closed, began to sip the water.

After Zea turned her head away, her mother wetted the towel and began to wash the dry blood from the girl's body gently. When she finished with the washing, she put a sheepskin over the girl to cover the lower part of her body.

"She's a grown up now," Dorug mumbled to himself.

"Mam," they heard the girl's feeble voice.

"Yes, my darling," her mother answered.

The girl turned her head again to look at Bria. Her eyes were full of tear. "I'm so glad to be home."

Bria and Dorug sat down on the edge of the bunk.

"Pap," Zea said.

"Yes, my daughter."

"Pap, I escaped."

"Thanks God, my darling," the man said. "You're the first one to do that. No one has ever returned from the Korvags."

"I escaped," the girl repeated and turned her head once again to face the wall.

"Rest now," her mother said, putting her arm over the girl's shoulder.

Zea raised her voice by a shade. "I can't rest," she said trying hard to fight back her tears. "I can't rest."

Several of the curious onlookers were already inside the house.

"I must speak," continued the girl. "I must speak... before I die."

"You're not going to die," Dorug rushed to say. "You're home now, and you'll be all right. You'll see."

"The Korvags," Zea went on. "They don't kill the girls."

They were all holding their breath.

"They'd kill those who refuse to obey their orders... They'd also kill those who become weak or ill... Do you want to know what they do with us?"

"Go on, honey," her mother encouraged her.

Zea started sobbing. Finally, she regained her strength again. "Mam, I'll tell you... if you come closer."

Bria leaned closer. Her face was turning gloomy as she listened to Zea's account. When Zea finished, she stood up, shaking her head.

"Horrible" she said quietly. "Our daughters are sex slaves of the Korvag chiefs. Their top leader keeps a large group in his harem."

An older man stepped closer. "Do you mind if I ask your daughter some questions?"

Dorug answered: "Frugo, you're one of the elders. I'm sure you have some important questions."

"Zea," Frugo began, "can you tell us more about the Korvags?"

"I'll try," Zea answered. "What do you want to know?"

"How do they come across the swamp?"

"They have many tree trunks floating on top of the water beyond the reeds. They tied the trunks together with ropes. It can't be seen from this side. The reeds hide that from our eyes. Their horses must go slowly when crossing this bridge. On both sides of the bridge, where the reeds are, the muddy water is shallow."

"Have you seen their villages? Are there many of them?"

"I've seen only one village on the way to Dudvo... Dudvo is where Nezir lives. He's the big chief. He lives in the biggest stone house I've ever seen."

When he saw that Zea was exhausted from speaking, Frugo turned to Dorug. "We've been a peaceful people from the beginning of times... The Korvags take advantage of our peaceful nature... I don't think this is right, Dorug. They take our daughters. They took all of our horses. They forbid us to domesticate any more wild ones. They even forbid us to have large axes. We keep obeying..."

Frugo could not finish his last sentence. A few of young boys stormed into the house, shouting at the top of their lungs:

"They're coming! They're coming! The Korvags are coming!"

A bit later, several scary looking soldiers entered the house. One of them had a thick leather cap on his head, only his face was not covered. He raised the heavy stone ax in his hand and began to speak in a rough voice: "I'm Zurda. Nezir sent us." The tone of his voice was enough to frighten everyone.

"Nezir wants Zea returned!" the soldier continued. He waved to a soldier behind him who stepped to the bunk and lifted Zea.

The girl passed out.

"Stop," Dorug shouted. "You can't do this to us!"

"Shut your mouth!" Zurda thundered. "Not another word!"

"I won't let you take my daughter!"

"The consequence of your rebellion is death."

Before Dorug had a chance to move aside, Zurda hit his head with the blade of the ax. Dorug's skull split open, and the man collapsed. He was dead.

"Oh, no," Bria shrieked.

The soldiers did not care. They took Zea outside, got on their horses, and stormed away.

Ravar kissed Lella's forehead, then he took his arms off her and stepped to the computer.

"We have some pictures coming in," he said. "Let's put them on the big screen."

"These pictures are moving at the speed of light," commented Lella.

"Not quite that fast... Let's see... I'll rewind them quickly... Now they are much slower."

"Perfect," Lella said. "Now I can follow... Look, shepherds! That guy is pointing at our Shuttle."

"We're a nice looking star there in their sky, aren't we?"

When they arrived at the scene where Dorug was killed, Ravar turned off the pictures. "That's where it's going to stop!" he said firmly.

"What are you going to do?" Lella asked.

"Well, it's time for our bioblaster. We'll show these primitive Korvags how strong they really are."

"Korvags... Korvags," Lella pondered. "What a strange name."

A young man, about his age, came to relieve Zuppo.

"Hey, Luum, what brings you here?"

"Go home, Zuppo," Luum said mournfully. "Your father is dead."

"What? No! Tell me it's not true!"

"It's true, Zuppo ... It's true. The Korvags killed him."

Zuppo did not wait another moment. He started running fast. He found his mother sitting quietly on the edge of the bunk she used to share with Dorug. Dorug's body was lying there next to her. The corpse was dressed into white canvas. His head was wrapped, only his pale face was visible.

Zuppo sat next to his mother and stayed there quietly for a long time. Neither of them said a word. Finally, Zuppo stood up and went outside. He found Frugo standing there by the gate.

"They're brutal animals," the elder said. "They have no mercy."

"This will have to change!" Zuppo replied, struggling to hold back his tears.

"They're three times as many as we are, perhaps four or five times as many. We have no way of knowing."

"Listen to me, Frugo!" Zuppo said with anger in his voice. "We've been sheep so far. Now we'll turn into foxes."

The old man had a troubled look on his face. "If we anger them, they can kill us all," he said.

"We'll have to fight back smart. Listen! While I was running home, a strange idea popped up in my head. I know they forbid us to have axes. Well, axes would do us no good anyway. What we need is a bow!"

"A bow, what's that?"

"It's something to force the Korvags to bow their heads."

"Are you daydreaming, Zuppo?"

"We'll have to find strong, young, thin trees, no thicker than my lower arm. We'll bend them and keep them bent with strong strings. We'll pull the strings, and shoot arrows right into the hearts of the Korvags. They'll keep falling from their horses like rotten apples."

The old man's eyes lit up with fire. "Let's go!" he said. "Let's talk to the men in the fields."

It took the Chroms only a few days to produce a bow for everyone, including those children strong enough to pull the string. They practiced shooting their arrows into the trunks of trees.

The next day, Zuppo and some of his friends got up before dawn to watch the new star shining over Mount Ulvi. After the star set behind the ridge, they talked about their plan to defeat the Korvags.

"How come the Korvags never thought of making arrows?" asked one of them.

"Come on, Morva! You know how ignorant they are," answered another one.

Someone else was wondering about the swamp. "What if they dismantle their bridge before we ever get going?"

"Why would they do that?" Zuppo replied.

"I have an idea!"

"Will you tell us about it, Zorel?"

"Let's dip the tip of the arrows into snake poison!"

"A very clever thought," replied Zuppo. "It appears to me that we are getting smarter fast. Now we just have to learn to get rid of our fear."

"Is it fair to say that the seven of us are now the leaders of the Chroms?" Zorel asked while turning to Zuppo. "Chroms, right? Is that how your Pap called us?"

"That's right Zorel, Chroms... As for being the leaders, we'll see. Our ability to think more freely, to come up with new ideas and new solutions can certainly elevate us to that responsibility... I say, let's continue doing our daily chores as we always have and, if you all agree, let's meet here every day at dawn. We'll see what we can do to achieve a better future."

When the first rays of the sun peeped out, they left their gathering place. Some of them went to the fields to take care of their crops, others to work on repairing houses. Zuppo had to bury his father's corpse.

They gathered again the next morning.

"I had some strange dreams!" one of them reported after they settled down in a circle. "I would have never thought I'd be able to conceive such dreams."

"I did, too!" another one said.

"Well, it seems like we are destined to make some changes," Zuppo said. "As for all that we talk about here... I think we should keep to ourselves, at least for the time being. Of course, everything we do, we do for the good of all Chroms. This doesn't mean that everyone has to know all of our thoughts."

"Agreed... agreed," they all said.

A few days later, Zuppo asked the elders to call the village together. They met just after sunrise, out in the open field.

"First of all," Frugo began addressing the crowd, "I'd like to tell all of you that, we, the elders, concluded that our people need a new leader. Zuppo, son of our late leader would certainly be worthy of our trust. However, he suggested that his whole team, the Sevens as we now refer to them, take the lead. We agree. From now on, they will make all decisions. We, the elders, will still be available when needed, although, Zuppo and his team appear to have a much better understanding of how to go on from here. Raise your hand if you disagree."

No one objected.

"Our weapons are ready," Zuppo took over. "We all know how to use them. Tonight, a select group of the men will go across the swamp. I'll lead them. My Mam told me what my sister told her about the Korvags' villages. Nezir, their high leader lives in the second village, not too far from the swamp. We'll take his house by surprise and free as many Chrom girls as we can."

"Kill Nezir!" someone shouted.

"We will, if we have to," Zuppo continued. "Our plan is to capture Nezir and bring him back with us. Perhaps we can trade him for the rest of our sisters."

"Good plan! Good plan!" the villagers agreed.

When night fell, two dozen strong, young men, the Sevens among them, marched to the swamp. They waded across the knee deep mud. They fought their way through the reeds. Their eyes quickly adjusted to the darkness. They found their way around fairly well in the faint light of the stars.

They reached the end of the bridge in knee deep water, not far from where the reeds thinned out. It was not easy to step from one trunk to another as there were big gaps between them. Zuppo was wondering how the horses managed, even in broad daylight.

Finally, they reached the other side of the swamp.

They did not have to go too far before they saw the houses of the first village. There was candle light in one house. Otherwise, everything was dark and quite. They kept a safe distance as they lurked on.

"This is it, I think," Zuppo said when they arrived at the second village.

They all got down to the ground and crawled to the stone wall that surrounded the houses. Staying close to the wall, they approached the gate. Zuppo and Zorel went ahead. They tiptoed to the gate where they saw two guards under the dying light of a torch.

"They're either drunk or dead," Zuppo whispered.

They sneaked up on them and slashed their throats with their sharp blades.

Zorel moved the dead bodies into the field while Zuppo went back for their comrades.

Inside the village, they came across a very large house.

"I think this is where Nezir lives," Zuppo said.

They found a big pile of dirt and hid behind it for a while.

"I'll be right back," Zorel whispered. When he returned, he carried a dead man. "This soldier was guarding the entrance to the house," he informed the group. "He, too, was asleep. The Korvags must've had a feast and drank too much. There're four other guards in the yard. They don't seem to be too alert, either. A torch provides some light and that's good for us."

"Let's split into smaller groups," Zuppo proposed.

All four guards were soon out in the field with their throats cut open.

As he sneaked carefully further in, staying always by the wall, Zorel heard someone snoring. 'It's coming from the other side,' he thought. He looked for an opening in the wall. As he looked through the hole, he could not believe his eyes. He saw a group of the girls from his village lying there in a large chamber. They were all on the ground, one next to another, sleeping on straw.

There was a torch over the entryway, opposite from Zorel's position. The dancing flames lit up the chamber, only the corners were in darkness. A soldier was partially blocking the entryway with his body as he was sitting on the ground, with his legs stretched, leaning his back against the wall.

It was the soldier who snored.

Zorel wasted no time. He found the entrance and quickly cut the guard's throat. After that he jumped to the nearest girl, put his palm over her mouth and woke her.

"It's okay," he said when the girl opened her eyes. "Don't be frightened! I'm here to take you home. All of you." He turned his head so that the light fell on his face. "I'm Zorel. Do you recognize me?"

The girl nodded.

Zorel withdrew his hand. "Wake everyone up," he whispered.

The girl immediately began shaking one of her neighbors.

When they were all up, Zorel, with the dead soldier on his shoulders, led them out of the chamber. Outside, they heard horses clapping their shoes against the stony ground.

"Don't worry!" Zorel told the girls behind him. "There are more of us here tonight. I'm glad someone found the horses."

When they arrived behind the pile, Zuppo's team had already returned.

"Well done!" Zuppo praised Zorel who threw down the body of the dead soldier. "How many girls did you find?"

"I counted more than two dozen... What's this weird sound?"

"You mean this?" Zuppo said and kicked a body on the ground. "It's the great chief of the Korvags moaning. He can't express himself any better as his mouth is stuffed with grass. Right big Nezir?" He kicked him again. "We tied him up, he can't even move."

The rest of the team arrived.

"We have about thirty horses," reported Morva.

"Very good," Zuppo said. "Everyone will take a girl on horseback. I'll take Nezir." He threw the Korvag on a horse with his legs hanging on one side, and his head on the other.

They quietly led the horses out of the village.

"Let's go way around that other village," Zuppo gave the order as they began their ride.

The horses had a hard time crossing the swamp. The men had to get off and patiently lead the animals through the bridge. Finally, they were back on their land.

The sun was about to rise as they approached their village.

A lot of the villagers were waiting for them at the stone wall.

"Welcome back to Tolup," Frugo greeted Zuppo and his men at the gate. "I see your mission was a success."

"Yes, Frugo, I think it was, indeed."

The men helped the girls off the horses.

Families were very happy to see their loved ones returning home.

Zuppo and Zorel held a quick meeting.

"Send the men home, Zorel. They need a good rest. Let them each take a horse. Tell them to have a new group ride back here as soon as possible. The Korvags could be on their way already. We'll have to go back to the swamp and shoot them with our arrows as they emerge from the reeds. That's probably our only chance to remain victorious."

"What about Nezir?"

"I'm glad you asked... Frugo!"

Zuppo pointed at the prisoner who had been lying to the ground. "We'll put him back on a horse. Can you take him to your house and have the elders get him interrogated?

"We'll do, Zuppo, we'll do," the old man answered. "I'm sure we'll have plenty of questions."

Morva, Grog and Zorel stayed with Zuppo. The rest of them went home.

"I'm afraid, we don't have much time to waste," Zorel said.

"You're right," replied Zuppo. "Let's go! However, one of us should stay behind and bring the fresh men when they arrive... Grog! You suffered a wound during the night. Why don't you stay here? You know what? Don't even come with the men. Just tell them to follow us and then go home and take care of yourself."

"I'll do as you suggest."

The three men galloped back to the swamp. They arrived just in time. One of the Korvags was already out of the reeds.

Morva was the first to grab his bow. He pulled the string and sent the arrow flying.

The Korvag soldier had no chance getting out of the way. He could not even cry out. The arrow went right through his throat. He tried to pull it out with both his hands but a little later he fell off his horse and landed face down in the mud.

"It's my turn," Zuppo said when the second Korvag showed up.

The enemy soldier died instantly when Zuppo's arrow penetrated his heart.

Zorel aimed at the third one. As the Korvag tried to duck, he got the arrow into his left shoulder.

"Go back!" shouted the Korvag to warn his people in the reeds. "Go back!"

He could not shout the third time. Zuppo's arrow finished him off.

"I'll get the horses," Zorel said.

"I guess the rest of them changed their minds," commented Morva.

"This saves us for a while," Zuppo said. "We'll have to station a dozen of our shooters here. None of the Korvags will make it across alive."

They did not have to wait too long for the new army to arrive.

"Not all of you will be needed," Zuppo told the men. He selected the guards and then sent Zorel and Morva with the rest back to the village.

The people of Tolup were in a good mood all day. Work was put on hold as everyone celebrated.

Zuppo later arrived at Frugo's house where he found Nezir in the company of five of the elders. The Korvag chief was tied down on top of a long table with the elders sitting around him.

"You will die!" the Korvag barked at Zuppo. "All of you! You will all die!"

"I'm afraid you're no longer in the position to decide our fate," Zuppo answered calmly.

"He won't speak to us," Frugo informed Zuppo. "He just keeps shouting all sorts of abominations and threats."

Zuppo turned back to Nezir. "Surely, you'll tell us where your men hide the rest of the Chrom girls. If not..." He could not continue his sentence because the Korvag chief broke into a roaring laugh.

"What did you call those whores, the Chrom girls?" Nezir's laugh changed to coughing as he was choking on the grass that remained in his mouth. He spat out a bunch of muddy grass and then he closed his eyes.

"Getting exhausted... Finally," noted one of the elders.

"Well, if he doesn't want to talk, it's okay," Zuppo said. "We'll keep him tied down until he changes his mind. I'm sure he'll ask for water and food later. We'll quench his thirst after he tells us what we want to know."

Nezir made an attempt to sit up but the ropes were holding him down firmly. "Listen to me, you bastard!" he shouted at Zuppo. "My soldiers will come as soon as my people discover I'm missing. They'll eat all of you alive. You'd do much better letting me go. Perhaps I'd have mercy on you then. Otherwise, you won't see another day. You have my words on that."

"Your soldiers already have come," Zuppo replied quietly. "Those who made it across the swamp are dead." He stepped out of the house and returned with the head of a Korvag soldier. "Open your eyes Nezir! Look here! I bet you recognize one of your own." He held the head close to Nezir's face.

A few drops of blood landed on Nezir's forehead.

"Zurda!" the Korvag yelled out. That's all he was able to say. His facial expression took a very grotesque appearance as he closed his eyes again.

"The rest of your men were wise enough to stay over on the other side," Zuppo continued. "We don't expect to see them again anytime soon."

"You're the devil!" Nezir said with his huge teeth clenched.

"You can call me anything you want," Zuppo replied. "It won't change the fact that your men killed my father." He tossed the bloody head onto the table and squeezed it against Nezir's face.

"Eh!" the Korvag moaned in disgust and turned his head as much as he could.

"What? Don't you like your own kind?" Zuppo lifted the head and held it over Nezir's face again.

"Take... this away!" Nezir shouted. "Then... I'll speak."

Zuppo took the head outside.

"The girls are scattered around in the villages," the Korvag started. "I doubt you'd find them on your own... Besides, another raid would certainly fail. My men would be waiting for you... Let me go and I'll send your girls home."

"Sounds so simple," Zuppo said. "The problem is I don't believe a word you say. Why would you keep your promise?"

"You have no alternative. You'll never be able to free them without my help."

Zuppo gazed outside through an opening in the wall. He had no idea of how to proceed. He was too tired to think.

"Go home, Zuppo," one of the elders said. "Go home and sleep. We don't need to find the solution right at this moment."

"You're right. I need to feed myself as well. However, before I go I have one more question for this murderer... Nezir! That's the name you listen to, isn't it?"

"That's my name," the Korvag answered, stretching the words.

"I need to know what happened to my sister, Zea! She's the one who escaped from your harem. She's the one your man, Zurda, returned to the other side of the swamp. She wasn't among those we freed last night."

"I gave her to Zurda, as a gift," the Korvag said after a pause. "Now, that Zurda is dead, she'll be inherited by another man."

"Who's that other man?" shouted Zuppo.

"You wouldn't know him even if I told you his name."

Zuppo's hand came down hard on the table, right next to Nezir's head. "I don't care whether I'd know him or not. I want his name!"

The Korvag took his time. "Sobor," he said at last.

"Sobor..." Zuppo took a deep breath while trying to remain calm. Both his hands clenched, he walked over to the other side of the table. "Where can I find this Sobor?"

"He lives in the third village from the swamp."

Zuppo stormed out. Outside, he sat on one of the stepping stones and buried his face in his palms. He knew it would be suicide trying to go over the swamp anytime soon.

He went back inside the house.

"Frugo, can I talk with you?"

The elder followed him to the yard where they both sat down.

"How long would it take to ride around the swamp?" Zuppo asked.

Frugo shook his head. "Don't even think about that, Zuppo! The swamp stretches out a very long way and in either direction you go it blends into impenetrable jungle. Those who dared to enter the jungle were surely eaten up by wild animals. No one has ever returned. According to legend, the animals there are huge, horrible monsters."

"Those jungles surely have an end somewhere!"

"Those woods are never-ending, Zuppo. Take my word for it."

"Okay, Frugo." Zuppo stood up. "I'll see you later. Make sure Nezir stays alive. We'll need him."

At dawn the following morning, the Sevens arrived at their gathering place just before the Brius reached the ridge of Mount Ulvi.

"Thank you all for the heroic accomplishment the night before," Zuppo opened the session. "I believe we taught the Korvags an important lesson. Of course, our victory is far from complete. Next, we'll have to free the rest of our sisters in captivity. Beyond that, our goal is to have lasting peace with the Korvags. The alternative is to continue fighting, perhaps for many generations to come. We certainly don't want that! We're not a warring people."

"What if the Korvags want to fight?" Zorel asked.

"I'm confident we'll find a way to change their minds. In fact, I'm already into something." He paused. "You may have a hard time relating to what I'm about to tell you... In my dream, just before I woke up this morning, a creature, whom I never saw in my waking hours, showed me something really amazing."

"A creature," Morva interrupted him.

"Well, a man... like you and I... but with an alien face."

"Go on, Zuppo!" another one said impatiently.

"This alien... This man taught me something valuable. I can now record anything we say, anything at all."

34

"What are you talking about?" a couple of the men asked almost simultaneously.

"I'll show it to you in the daylight. I'll need sheets of leather or canvas. I'll cut the stem of a feather lengthwise and dip its pointed tip into black walnut tree extract and make my markings on the sheets. The extract will dry and people will be able to read the markings."

"What markings?"

"I'll show everybody! All those sounds that make up what we say will have distinctive marks. These marks put together will record everything we can possibly think of. You'll see how easy it is."

"You already know how those marks look like?" Zorel asked somewhat skeptically.

"That's right, Zorel." Zuppo pointed at his own head. "I don't know how it happened but they're all in here. I can see them clearly if I close my eyes."

"You got me curious," Zorel said.

Grog, who had some white clothes wrapped around his head to cover his wounds, picked up a piece of a dry branch and handed it to Zuppo.

"Here," he said, "take this! The sun is rising, now we have enough light. Can you make some of your markings right here in the sandy soil?"

Zuppo took the branch and carved out four separate signs, two of which, the first and the last ones, were identical. He pointed at them one after the other as he slowly said: "G-R-O-G."

"My name?!"

"Could you do mine, too?" Zorel asked.

Zuppo carved out the other six names: Zorel, Morva, Tudos, Bodor, Toma and Zuppo.

"This looks really interesting" Toma said pondering. "However, I don't see how it can help us achieve peace with the Korvags."

All eyes turned to Zuppo.

"Well, right at this moment, I'm not sure I can see it myself," Zuppo replied. "All I know is that recordings will play a role."

"When will you show us everything?"

"I'll have things ready for tomorrow morning... Now, what we should talk about is Nezir, the Korvag chief."

Grog raised his hand. "I have an idea."

"You want to kill him," Toma said.

"No, I don't think we could afford doing that," replied Grog. "Instead, what we have to do is this... Let's take him back to the swamp and have him shout over to the other side. Now, I'm sure there are some Korvags who stand guard there."

"Not bad," Zorel said, "if Nezir is willing."

"We'll make him willing" Zuppo replied.

"What will he have to shout?" Toma asked.

"That's exactly what we have to come up with," replied Zuppo. "We might even leave it up to him. After all, he knows his people. We just have to make it clear to him that we want our sisters returned. He probably has some ideas about this already... I'd bet he doesn't tremendously enjoy being tied to a table in Frugo's house. In fact, he told me yesterday that if we let him go back, he would send home the girls. Grog's idea could very well work."

"Let's go and try it!" Zorel proposed.

They all agreed.

Nezir was asleep when they arrived at Frugo's place. Still tied down, the Korvag chief lay in a puddle of smelly liquid.

"Don't despair, Frugo," Toma said, "we'll wash your table clean when this is all over."

"Nezir, are you ready to cooperate?" Zuppo asked after the Korvag opened his eyes.

"I'm ready," Nezir answered in a deep, hoarse voice.

"His throat must be awfully dry," Frugo explained. "I didn't give him any water during the night. I didn't want him to urinate any more."

Zuppo and Zorel took off the ropes that tied the Korvag to the tabletop.

"We won't untie your arms yet," Zuppo said. "It's just safer that way."

The Korvag nodded as if he was taking Zuppo's words as a compliment.

"However, we'll take the ropes off your legs. You wouldn't survive another trip hanging from a horse... I warn you, Nezir! Any attempt to escape, and you're dead. My arrow will pierce your heart. Can you ride with your hands tied?"

"If I have to," replied the Korvag. "Where are you taking me?"

"Back to the swamp..."

"That's what I thought. You'll want me to shout over to the other side?"

"You're smart, Nezir!" Zorel said. "You're very smart. I guess you've also figured out what you'll have to shout to your men."

"You want them to bring the girls."

"Exactly..."

"When you have the girls... then what?"

"You'll be free to go."

"Will you keep your word?"

"You only have one way to find out, Nezir."

Frugo stepped closer to the Korvag. "We also want you to promise something," the old man said. "You and your people will leave us alone in peace."

"We'll do," Nezir replied after some thinking.

When they arrived at the swamp, the Korvag chief yelled out several names. Finally, someone shouted back from the other side:

"What's happening, chief?"

"I want you to go and gather all the girls that are from Tolup. Bring them here! You'll move them over the reeds on your horse one by one."

More than half a day passed before they heard the voice again from the other side:

"Chief, are you still there?"

"Did you bring them?" Nezir shouted back.

"They're all here. I'm coming with the first one."

When the horseman, with a Chrom girl sitting behind him, emerged from the reeds, Nezir yelled at him: "Stop right there!"

The Korvag chief turned to Zuppo. "I know you're anxious to see your sister, Zea. I'm just as anxious to be home." He spoke loudly so that the horseman could hear him as well. "Zea will cross last, after I returned. My man will stay here to take the horse back."

Zuppo objected: - She won't be able to ride alone!"

"I'll go over with Nezir," offered Bodor, the quietest of the Sevens. He then raised his hand, as Zuppo was about to argue. "Zuppo, I beg you! Let me do it! You know how much Zea has always meant for me."

Zuppo knew it was Bodor's aching desire to bring Zea home. So, he nodded and then turned to Nezir. "Let me cut the rope off your hands."

Once his hands were free, the Korvag chief waved to the horseman to proceed.

During the whole time, the Chrom guards stationed there were standing by with bows in their hands, ready to shoot in case something suspicious happens.

Finally, all the girls were accounted for, with the exception of Zuppo's sister.

After Nezir and Bodor disappeared in the reeds, the passing of time became painfully slow, especially for Zuppo. 'What if Nezir kills Bodor and takes off with Zea?'

At long last, the head of the horse appeared and then they saw Bodor and Zea. The girl was waving toward them.

After Bodor and Zea got off, they sent the Korvag back on the horse.

While Zuppo and his people were heading home, and Nezir was riding back to Dudvo, Wrix decided to take a break and have some holographic fun. In a moment, he beamed himself to the jungle.

"At last!" he said loudly after making himself visible and activating his voice generator. "What a difference it makes when one is free."

He turned off the primary link between himself and the host computer in the Shuttle.

"I could actually spend the entire afternoon here. The pause in the transmission for Lella and Ravar would appear very short. They would not even notice it."

He enjoyed hearing his own voice.

Looking around, he saw giant trees and lush green vegetation. The soil was muddy, and there were patches of water here and there. As he had no body weight, he did not have to worry about sinking.

"Nevertheless, I'd much rather stand on dry surface." As he said that, he beamed himself further away from the swamp.

"Now, that's more like it."

He looked up and saw fragments of the blue sky. The rays of the sun broke through the branches, straight into his eyes. He smiled. He knew that the bright light would bother any living creature's eyes. It did not bother his at all.

The trees at this location were much shorter, and they branched out close to the ground. The soil was reddish and sandy with thin, dry grass.

"What the heck!" he blurted in surprise when he saw a furry mass flying right through him.

It was a large, striped beast that jumped on him from the branches. It had its enormous jaws wide open, set to tear Wrix's head off. It roared as it sailed through the air.

The animal was equally surprised when it landed on the ground in an awkward position. It was expecting to take Wrix's body down which, of course, did not happen.

Wrix watched the beast rolling over before getting back on its paws. He found it very amusing. He laughed wholeheartedly.

The animal took up an attacking position, opened his jaws and roared.

Mimicking the animal, Wrix got down on his knees and elbows, opened his mouth and howled back.

The animal sprang off the ground and landed in Wrix's hologram. It locked its jaws and jerked its head from side to side. Apparently, it was trying to tear Wrix apart.

Wrix suddenly beamed himself to a branch above, took up a sitting position, and continued laughing at the foolish animal.

When he had enough of this game, he got on the back of a soaring bird to enjoy the scenery below.

Back on the ground, a long, bulky snake tried to harass him. The reptile attempted to bite him several times, releasing its venom into his legs and arms. Of course, the poison went wasted, getting soaked up in the ground.

Wrix teased the snake for a while and then he moved on. He was now looking for a place where he could spend some time undisturbed. When he found his spot on top of a rock, he lay on his back and stared at the blue sky. He was wondering what good it would be having a real body just to be torn apart by some merciless beasts in the jungle.

"I am so glad I'm just a hologram," he muttered. "On the other hand, even a short lived real existence might be more meaningful than being a potentially everlasting image made up of photons and electrons."

Of course, he knew that a real life for him was out of the question.

"Then why did my creators program me to have the comprehension of what real life would be?" He paused and then he went on muttering. "On the other hand, knowing is better than not knowing... Besides, my creators probably tried to make me as perfect as possible. Obviously, they could not produce an organic body for me. They did what they could.

They loaded all their mental and emotional experiences into my circuits. Even though I can not be one of them, I can perceive the world as if I was."

He sat up and looked into the distance over the top of the jungle. He saw the rising mist in the distance, the patchy white clouds in the sky, the rays of the sun reflected by water trapped in huge leafs on treetops. He saw the birds. He heard the wind whistling through a hole in the rock.

"It's all so beautiful," he sighed.

As much as he wished he could be truly free, he knew he had no alternative to being a slave. That's what he considered himself at this moment, a slave.

"In a way, it's outrageous," he said. "Should I decide to escape or even to disobey, Ravar simply turns off a switch, and I cease to be. As soon as the energy stops flowing, my thoughts are non-existent."

A horsefly tried to land on his nose but it just could not find a solid platform.

Wrix was entertained by the many attempts and finally he started laughing again. "Wow! These fine, silky, vibrating wings would surely tickle me to death if I was of a fleshy substance," he joked when the fly entered his nostrils.

Suddenly, one of his hands swooped through his holographic head to catch the little buzzer. For a while, he watched the fly struggling to free itself. Finally, he opened his hand and said: "Go! Be free. I have much to do."

After the fly took off, Wrix re-established his link to the host computer and rendered himself invisible.

Nezir lay dead on top of the pile in front of his house. The blood that had flowed from the hole in his skull dried on the soil.

The sun was about to set.

A group of Korvags stood around the pile. They were listening to a young, muscular soldier who still had in his hand the bloody ax that ended the big chief's life.

From time to time, while the soldier spoke, someone from the group stepped closer to the pile and spat on the dead body.

"What a disgrace," thundered the soldier, "he betrayed all of us. A coward is what he was. Instead of dying with dignity while in captivity, he became a traitor. He sold us out. He returned our precious possessions just to save his own miserable life. He had no idea of what it takes to be the chief of the Korvags." He paused. He carried his glance around to see if there were any comments.

"Every coward and every traitor will die! I, with my own hands, will carry out this sentence." He raised the ax in his hand. While looking at the dry blood on the blade, he added: "Any objections?"

The men looked at one another.

"Sobor, you're our new chief," one of them said.

"You're the chief!" another one shouted.

"Sobor is the chief!" they all shouted.

Sobor threw his ax on the ground.

"I trust the wisdom of your decision and I'm honored by it," he continued in a softer tone. "I'll lead you and the rest of the Korvags to a happier life. All of you will get what your hearts desire. We won't deny ourselves what is rightly ours. You have my word on that."

"What about the girls from Tolup?" one of the men asked while Sobor paused.

"That's a very appropriate question, Bruvo. Step out and come, stand here beside me."

Bruvo, another heavy-built, young soldier, stepped right next to Sobor.

"I like courageous thinking," said the new chief, and he briefly placed both his hands on Bruvo's shoulders. "You're my right hand man from now on. Let me see your ax!"

Bruvo handed the weapon to Sobor. "It's very sharp, chief," he said proudly. "I always keep it that way."

"Very good," Sobor acknowledged. He turned back to the group. "Forget all that gibberish Nezir spoke about peace with the Chroms. The only form of peace I accept is their total submission. We'll conquer them once again. First of all, our horses must be returned. Of course, the Chroms will have to continue sending us half of their crops." He glanced at Bruvo. "The girls we've been used to must come back to our houses."

"What about those poisonous arrows?" someone from the group asked.

"I was about to address that issue!" Sobor replied somewhat annoyed. "We're no dumb. We can make those string weapons, too. We'll shoot them. We'll shoot them all if we have to."

"How are we going to enter their territory without getting shot first?" inquired the same man.

"I suppose you dare to ask these questions because you think you have better ideas than I do," Sobor snapped back angrily.

"Not at all, chief. Not at all," the questioner rushed to explain. "I'm just worried."

"Don't be worried!" Sobor shouted. "We'll defeat them... Now I want those of you who were Nezir's commanders to step forward!"

About a dozen soldiers, all of them with leather caps on their heads, separated themselves from the group.

Sobor looked at them one by one and then he said: "Those of you that pledge alliance to me raise your hands! Good. I see none of you hesitated. You all will be next in command after Bruvo. Now we go and feed ourselves. I want all of you, commanders, be back here when the stars flare up. We'll work out our strategy. I'll need your clear thinking so save the wine until after our meeting."

Bruvo quickly raised his hand. "Chief!" he shouted. "Do you want me to bring Torza? He's still out there on the bridge with the rest of the guards. You know, Torza is the one who returned the girls."

Sobor thought for a while. Finally, loudly, so that everyone could hear him clearly, he said: "We'll let Torza live. After all, he was only carrying out Nezir's order."

Zuppo's demonstration fired up the Sevens.

All of them were eager to learn how to make sense of the markings and record them on sheets of canvas.

By the time the sun reached its highest point in the sky, they all have mastered the art of writing.

"You can now all go and teach everyone in the village. Tell the people to record anything interesting that happens in their lives." Zuppo then looked toward Mount Ulvi. "I had a dream last night," he continued. "The man who taught me the art of recording appeared to me again. He told me to pack bread and water, and go to the top of the mountain as soon as possible."

The others looked startled.

"Why?" Zorel asked. "Why go there?"

"It's dangerous up there," Morva said. "You know the volcano is active in Mount Ulvi. It can erupt anytime without warning."

They all said something trying to talk Zuppo out of his planned journey.

Zuppo smiled.

"Zorel," he said after listening to all the reasoning, "I'd like you to take the lead while I'm gone. Conduct the meetings at dawn as usual. I now have to go and prepare for my odyssey. Farewell to you all."

Zuppo went home and packed his bag.

"I don't know when I'll be coming back," he told his mother and his sister. "Take care of each others."

Zea had tears in her eyes when Zuppo left the house.

"He'll be all right," her mother consoled her. "Don't worry, he'll be back."

They did not talk about it but they both knew Zuppo had changed. They also knew his dreams changed him.

Bria realized her son was now more than just an ordinary man. The changes took place under her watchful eyes, and they all happened so suddenly. All she could do was pray.

Zuppo walked tirelessly the whole afternoon. His moccasins were soft. The leather protected his soles from the rough surface of the ground. The airy pants and shirt his mother made of canvas were comfortable. He had a sheepskin rolled up and tied to his lower back with a rope. It was to save him from the cold of the night.   His bow hung from one of his shoulders, his bag from the other one. His sharp blade in its leather holder was fixed to his belt.

It was hot under the scorching sun, he sweat profusely. Still, he only drank small portions of his water. He knew he would not find a spring until he reached the foot of the mountain, and that was rather far away.

When night fell, he rolled out his sheepskin under a shrub and went to sleep.

It was still dark when he woke up. The first thing he saw as he opened his eyes was Brius, the star he had discovered.

'It shines so much brighter now,' he thought as he watched it pass behind the ridge.

He allowed himself a small ration of his bread, a sip of the water, and then he continued his journey.

All day long he was thinking of his people. He knew the Korvags would eventually find a way to fight back. His sister, along with many other girls, would be captured again and taken back to the other side of the swamp.

'We are so badly outnumbered. How could we possibly defend ourselves?'

His brain worked feverishly.

"It's so unfair!" he said loudly. "It's so unfair! It should be the other way around. If we were in the majority, the Korvags wouldn't dare to bother us. We sure wouldn't bother them. We could all live in peace."

As the sun turned deep red over the horizon, ready to bring the end to another day, Zuppo reached the foothills.

He spent the night under a cliff, right next to a wall of stone. Although he heard the barking of wild dogs in the distance, he felt safe.

After the exhausting day, he slept deeply all night long.

His third day was more treacherous.

Further up on the mountainside, he ran into all sorts of challenges.

First, he had to struggle through very thick shrubs. The thorns cut open wounds in his skin. His clothes became ragged.

Later, high cliffs blocked his way. Anywhere he looked, the rocks were steep. There was no other way, he had to pull himself up with his hands. It was a very slow, arduous ascend.

About half way up on an almost vertical wall, it seemed like there was no way getting any higher.

'Turn around and go home,' he heard a voice inside his head.

"No!" he cried out as loudly as he could. "I won't turn back!"

He felt sharp pain as the edge of the rock cut into his palm. He sensed the warm blood running down on his arm, over his shoulder, and then onto his back.

'It's not late to change your mind,' he heard the voice again. 'You can go down the same way you moved up this far. You can still save yourself.'

"No!" he shouted. "No! No! No!"

Giving it all he had, he swung to the side and caught the stems of some strong weed. He pulled himself up, and finally he found support under his feet.

He rested for a while and then he climbed to the top of the cliff.

On his feet again, he fixed his eyes on the very top.

"Whatever it takes," he muttered.

In that moment, he felt sharp pain in his lower leg. He looked down just to see the snake bite once more. He grabbed the reptile and threw it over the cliff.

He knew he had no time to waste. He pulled his blade and cut into his leg to remove a chunk with the venom in it. He did not care about the pain. He pressed one of his palms against the bleeding wound while his other hand was busy collecting leaves. He used whatever water he had left to wash out the wound and then he placed several layers of the leaves over it. Finally, he wrapped his belt around his leg to keep the leaves in place.

'What else?' he was wondering as he looked at his still bleeding palm.

Once again, he glanced at the peek.

'It might take another day or two,' he estimated.

The sun was right above him, and he had no more water.

"There!" he cried out loud when he saw the forest in the distance. "There must be water in those woods!"

His leg was throbbing with pain as he walked on. Fortunately, the ground became fairly even and the slope not too steep.

It took him half the afternoon to reach the edge of the woods. He went to the first tree and settled down to rest in the shade. Leaning his back against the trunk, he closed his eyes. He fell asleep in an instant.

A shrill sound woke him up. It came from a large bird that had just landed on a branch above him. He reached for his bow and sent one of his arrows through the bird.

He cut off the head of the bird with his blade and began to drink the fresh blood. He gulped several times but then he threw the bird away. Suddenly, he had to vomit.

'Water, I need water!'

He felt sick. His whole body was in pain.

Out of nowhere, a snake appeared. It held its head high as if ready to bite. However, it did not attack. Instead, it opened its jaws and began to talk: "Go home, Zuppo. Go home. You don't belong up here. The volcano will swallow you up. You will surely die. It's still not too late for you to turn around."

A moment later, the snake was gone.

Zuppo was lightheaded. He touched his face. It felt very hot.

"Delirium," he mumbled. "There's no snake... It's the fever."

He managed to get on his feet and went deeper into the woods. All of a sudden, he stopped. He thought he heard a waterfall.

He was right. Soon, he found a small lake fed by a stream that came from the side of a rock. The water in the lake was crystal clear.

He leaned down and drank.

After he quenched his thirst and filled his leather bag, he found his way back to the edge of the woods and resumed his course to the top. As he walked, he ate half of the bread he had left.

His strength returned as his fever subsided.

'Well, one more night,' he thought when it turned dark.

He found a lonely tree with thick branches. He climbed up and settled in for the night. He was in no danger of falling down in his sleep as two of the branches that grew very close held him securely. Of course, it was not a comfortable way to spend the night.

In the morning, he was mentally refreshed but his body was aching. He had to walk quite a distance before he began to recover from sleeping on the tree.

The wound in his leg was still sore but it felt much better now. His fever was completely gone. If any of the snake's poison got into his bloodstream, his body had already neutralized it.

"By tonight, I must be up there!" he said resolutely. "Nothing will hold me back."

His sentence was finished by a deafening thunder. First he thought the volcano was erupting but then he saw clouds rushing from behind the mountain.

The wind started blowing as the sky darkened. Lightening zipped back and forth above his head, creating one continuous, awfully loud thunder.

He had no choice but to hurry back to the tree where he spent the night. By the time he got there, heavy rain started pouring down. The storm was getting stronger and stronger. The gusts tore off large branches of the tree and hurled them toward the sky.

Zuppo was struggling to hold on to one of the main branches that grew all the way down to the ground. He was hoping that branch would be strong enough to withstand the ferocious storm.

Ice began to fall, some as large as plums.

Zuppo had to place the sheepskin over his head for protection.

Lightening hit the tree and set it on fire.

Zuppo was thrown several feet by the impact. He hurt one of his ankles on landing.

He huddled under his sheepskin and watched the tree burning.

Soon, the rain killed the fire.

The storm left as quickly as it arrived. The sky was blue again, and the sun dried up the water from the ground. For a while, a layer of mist covered the mountainside, and a large rainbow appeared.

Zuppo examined his ankle. He thought it was nothing serious.

He resumed his course once again.

Late in the afternoon, he reached the highest peak.

Sobor and his commanders wasted no time. After showing their people how to make bows and how to fix sharp blades at the tip of their arrows, they assembled an army of volunteers. The army numbered more than the entire population of Tolup.

"We'll eat them alive if they dare to resist," Sobor told the commanders during one of their usual after supper meetings.

Bruvo suggested that they carry out their retaliatory attack during the night.

"I don't see why not," Sobor accepted the proposal. "The Chroms were able to cross the swamp in the dark. Why couldn't we do the same? I'm surprised we've never thought of a night crossing before."

One of the commanders also had an idea: "Let's send a few men to cut a passageway in the reeds. It doesn't have to be wide, just enough for the horses to move freely. The last few rows on the other side would be cut during the night, just before the attack."

"Very clever," Sobor acknowledged. "This will make everything simple. We'll be riding through fast. Their guards won't be able to see where to shoot." He paused before adding: "Tomorrow night! We're moving in tomorrow night. Bruvo! Round up a few men in the morning and have them cut the reeds. Tell them to do it quietly. Have them wait there to finish their work when we arrive."

Everything happened the way Sobor dictated.

The attack began shortly after nightfall.

The Chrom guards were quite surprised when the horsemen stormed out of the swamp. Some of them jumped on their horses and started shooting their arrows. Although, the night was not dark, most of their shots missed. When they realized they had no chance, they tried to flee. None of them was able to escape, they were all murdered.

The Korvag army surrounded Tolup and waited for the night to pass.

At sunrise, Sobor sent in the soldiers who took the village without much difficulty. Only a few of the Korvags died. Later, those Chroms responsible for the casualties were summarily executed at the village-square.

Many of the younger females were dragged to the square.

Sobor demanded those men who planned and led the raid against his people.

"These girls will die if your leaders don't show up right here, soon!" he shouted. He grabbed one of the girls by her hair and raised his ax above her head. He was about to demonstrate that he was not just fooling around.

"Wait!" It was Frugo who stepped forward.

"Who are you?" Sobor asked in surprise while letting go of the girl's hair.

"I'm the leader here!" Frugo replied.

"We're the leaders here!" Two other elders joined Frugo.

Sobor released a roaring laugh. "You, leaders... Next you'll tell me that you rode over the swamp to kidnap Nezir and to free the girls? Huh!"

"Chief of the Korvags," Frugo said firmly. This seemed to catch Sobor's attention. "Of course, we're frail, old men. You're right. We didn't take part in the rescue. We only planned it and organized it. No other Chrom is responsible for what happened."

Sobor suddenly changed tone. "Who freed the girls?" he barked. "Who seized Nezir?"

One of the Korvag soldiers stepped in front of Sobor. "Chief!" he said. "I can answer your questions. As you probably know, I was there when Nezir returned. I was the one who carried out his order to let the girls go."

Sobor looked at Bruvo and whispered into his ear: "See? I knew we had to keep this man alive."

"Okay, Torza, go ahead," Bruvo said, "name the bastards!"

"That's one of them." Torza pointed at Zorel who had just arrived at the square.

Zorel walked straight to Sobor and looked him in the eye. "Big, strong chief of the Korvags," he said rather mockingly. "Here I am. Let me bow my head so that you can easily cut a hole in it with your dreaded ax." His face turned angry as he began to shout: "Here I am. Kill me! That's all you and your people can do. Kill, kill and kill. That's all you can do." He quieted down as he shook his head.

"Where are the others?" Sobor asked calmly.

"Right there," Zorel pointed at the dead bodies of the Chroms executed earlier. "The rest were out there by the swamp. Those were the men who tried to guard our peace. You must have murdered them all."

"You got that right!" Sobor snapped back.

"I don't see the other one here," Torza told Sobor. "His name is Zuppo. He is the real leader."

"Where is that Zuppo?" Sobor roared at Zorel.

"You'd have to kill me before I answer any more of your questions," Zorel replied.

"Tie him!" Sobor instructed his men.

The soldiers knocked Zorel to the ground and tied ropes around his arms and legs.

"You're next, old man!" Sobor yelled at Frugo. "Where is Zuppo?"

"We haven't seen him for days," Frugo answered after some hesitation. "He left the village. No one knows where he is."

Sobor turned to his assistant: "Bruvo, take as many soldiers as you need and search every house!"

"You'll be searching in vain," Frugo said.

"If not, you'll die!" Sobor shouted at Frugo. "In the meantime, you'll be tied as well ... Go Bruvo, find Zuppo!"

The soldiers rounded up all the men in the village and brought them all in front of Sobor. Torza kept shaking his head.

"Sorry, chief," Bruvo reported later. "No more place to look."

Sobor pointed at Zorel and Frugo. "Throw these two on the horses! We'll keep them in our prison until Zuppo is ready to give himself up. If we don't have Zuppo a year from now, we'll kill them. Take all these girls with you, too. We'll distribute them when we get home."

The Korvag chief turned to the two elders and to all those remaining in the square: "I want all the horses brought here without any delay! Don't forget to turn in all your bows, either! Anyone who dares to hide one will be killed. Anyone who dares to make a new one will also be killed. I hope I made myself clear. As long as you don't try to cross the swamp again, and as long as you deliver us the usual portions of your crops, you can live in peace."

By the end of the day, the two prisoners as well as the three dozen Chrom girls were taken to the other side and all Korvags returned to their own domain. Some of the soldiers stayed behind at the bridge to guard the passageway in the reeds.

"This way, Zuppo... Come this way."

Quite startled, Zuppo looked around. He saw nobody.

"Come this way, Zuppo."

He had no problem establishing the direction the voice came from. However, he had no idea of who could have generated it. His skin began to develop goose bumps.

He touched his forehead to see whether he had fever again.

"This way, Zuppo..."

"This is no hallucination," he told himself loudly. "This voice is real."

"You're right, Zuppo, this voice is very real. It belongs to me. I know you can't see me but I'm right here."

Zuppo hit his head with his hand as if he was trying to wake himself up.

"You don't have to do that, Zuppo. You're awake and relatively well. I know your doubts would be gone if I could show myself to you. Unfortunately, I can't do that. You just have to rely on your imagination."

Zuppo realized what might be happening.

"Are you the one I saw in my dreams, the one who told me to come up here?"

"Right, now you know who I am."

Zuppo sat down. He decided he would not just blindly follow the voice. "I must tell you," he said, "that I had some weird experiences on my way up here. I even saw a snake that talked to me. So, how would I know you're really who you say you are? Forgive me for my doubts if you're for real but I need some kind of proof. I'm sure you understand."

"I understand you very well, Zuppo. Your doubt is a sign of your wisdom. However, you'll have to be able to overcome your lack of faith. You're about to embark on a mission to bring extraordinary changes in your people's lives. You'll be the great Prince of the Chroms. You'll deliver your people from the cruel oppression brought upon them by the Korvags."

"Excuse me!" interrupted Zuppo. "I'm only a simple mortal. If you really want me to overcome my doubts, you'll have to show yourself to me. I'll have faith as soon as I can see you. Until then, I'll have to assume that my mind is playing tricks with me."

Wrix did not immediately know how to proceed. Ravar instructed him not to take solid form unless it is really necessary.

'Well,' he reasoned, 'this situation certainly warrants disobedience.'

"All right, Zuppo, here I am!"

Wrix materialized right in front of Zuppo's eyes.

"Wow!" This was all Zuppo could say.

"Now you believe my voice is not only an imagination of your mind?"

Zuppo looked at the man in disbelief. "What strange clothing!" he wondered loudly. "What are those tiny flashing lights on your arms and on your chest? Otherwise, you look just like me."

"Really..." Zuppo's remark caught Wrix by surprise. He found it necessary to disappear for a moment. "I can fade in and out anytime," he said after appearing again.

"Are... you God?" Zuppo braved the question.

Wrix hesitated. "Well, in a way... yes."

Zuppo buried his face in his hands. It just occurred to him how rude and disrespectful his behavior had been. "Please, forgive my ignorance!" he begged after he threw himself on his knees. "I'm terribly sorry if I've offended you."

"Come on, Zuppo," Wrix said jovially. "You can't offend God! Unless you're God yourself and that is clearly is not the case here. Really, there's no need for you to be on your knees. As for your behavior, we know there's a reason for everything. Your imperfection is not your fault, and I'm not going to hold it against you."

Zuppo stood up, his legs shaking. He stared at Wrix with eyes wide open.

"You really are God," he whispered. "My God, I can't believe this is happening to me!"

"Well, now, Zuppo. You'll have to make up your mind!"

"Oh, God... You know what I mean. It's so incredible... Can I call you God?"

"You can call me whatever you like, that won't change a thing. Now, before all your hair turns gray, I'd like you to follow me down to that stony clearing. If your knees are still shaking, I can take you there."

"Oh, no, I'm quite all right!" Zuppo hurried with the answer.

Down at the edge of the clearing, there was a gap in a stone wall.

"There's a cave in there," Wrix said. "I hope you'll find it comfortable... Go on! Enter! This is where you'll stay for a while."

Zuppo inched his way through the opening. His knees were shaking again. He just had no idea of what to expect inside.

He found himself in a chamber. As he stood there, the stony walls all around were radiating a strange, greenish light.

"Where am I?" he asked as he turned around.

Wrix was not in the chamber.

Zuppo quickly went back to the terrace.

"God, where are you?" he shouted.

There came no answer.

Zuppo carried his glance around. 'The sun must've set already,' he thought as he saw the thickening darkness down in the crater.

He went back to the illuminated cave. Still in shock, he touched the walls to make sure they were real. His brain was working harder than anytime in his life before. He was trying to grasp all that had happened to him since reaching the peak but that was an overwhelming task.

"Miracle," he mumbled, "miracle!"

Eventually he realized he could not just stand there forever.

"What I really need now is a good rest," he said to himself.

He put down his sheepskin on a relatively even surface of the ground. He was not hungry but his thirst was almost unbearable. He reached for his water-sack and squeezed out the last drop.

'I'll die if I can't drink more very soon,' he thought. Just then, he heard water dripping. He found the tiny spring in a niche. He collected the water in his palms.

After quenching his thirst, he wrapped himself into his sheepskin and fell asleep.

It was getting dark outside.

Bria felt painfully lonely. She lit up a candle and sat down to eat her supper.

She broke up a slice of the bread she had baked earlier and placed the pieces on the plate her late husband used to eat from. The milk in the jug was still lukewarm as she had just milked the goat.

She only had a couple of bites. Her appetite was no longer as it used to be. These days she had almost no desire to feed herself.

She tried to imagine how Dorug would enjoy the fresh bread. She would tease him again about the milk on his mustache or the crumbs of bread in his beard. She would lean over the table and pick the crumbs one by one. Dorug would hold her hand in his huge palms and look at her gently, lovingly.

'It'll never happen again.' As she thought that, teardrops rolled down on her face.

Slowly, she put the pieces of bread into the sack. She drank some of the milk and then she placed the jug into an opening of the wall in the pantry where the draft kept it cool.

She extinguished the candle and sat in the dark for a while. She was in no mood to lie down. She knew she could not fall asleep.

Later, she stood up and found her way through the darkness out into the yard. Under the starry sky she could see more. She sat on the trunk of a tree Dorug pulled home on his cart shortly before his death.

She wiped the tears from her eyes with the back of her hand and looked up into the sky.

"Good God, somewhere among those many stars," she said quietly. "I hope one day you'll make me understand why things happen the way they happen. I lost my daughter the second time. I wonder whether she's still alive. If she is, how much pain does she have to endure for her young age? How long can she survive as a slave?"

"My beloved husband isn't coming home anymore. Why did he have to die? Why? He was still so young. How about my son? Am I going to lose him, too? Is he going to get killed just like his father?"

She had to swallow her tears.

"Why is it that we can't live in peace? We've never harmed the Korvags. Why do they harm us? Why? Our men are peaceful, gentle and gracious. Why are the Korvag men so brutal and so merciless? Are they ever going to change?"

"God, you're the only one who has the power to change them."

She heard the hens making some noise in their cage at the back of the yard. After they quieted down, she gazed at the stars again.

"God, I don't know whether you're listening to me. I hope you are. If there's a chance, please bring Zea and Zuppo back home. Please, let them live here in peace. Let them have families, children. Let them be joyful. They're so very young, they haven't even lived yet. Don't let their lives be taken away before their times come. If someone else from this family must go, let that be me."

Suddenly, she felt an urge to go back inside the house.

She was taken aback when she saw the candlelight inside. She could have sworn she had not left the candle burning. But when she saw what she thought was the ghost of a man, she almost fainted.

"Bria, don't be scared!" said the ghost. "Please, come closer and sit down."

The woman's heart was pounding.

"Who... What are you?" she stuttered when finally she was able to speak.

"I came to assure you that your son is all right. You needn't worry about him."

Bria cautiously moved closer to the table, slowly sat down and leaned forward, hoping to see more of the ghostly image. Instead, the man became even more transparent as the candlelight blinded her eyes.

"I must be dreaming," she said.

"What you're experiencing is real. This is not a dream. I came to you because I know you're in great pain. The loss of your husband probably wouldn't be so unbearable if he had died of natural causes. Let me tell you about your daughter as well. The Korvags aren't treating her very nicely but she's surviving the ordeal. I know losing her is another wound in your soul. Perhaps you get some relief from knowing that she's healthy and full of hope."

Bria extended her hand and tried to touch the man. When her hand moved through the ghost, her heart almost stopped. She quickly withdrew her hand and covered her mouth with it.

"No need to panic, Bria. I know you're extremely overwhelmed. If you want, I go away."

"No! Don't go!" she said quickly. "Please, don't go. I'll be all right in a moment... I'll be all right... What do you know about Zuppo?"

"He isn't coming back home anytime soon. He has to go through the transformation before he can return."

"Transformation..."

"He isn't going to be an ordinary man when you see him again. He won't even look the same. He's been chosen to be the Prince of the Chroms. He'll have great powers when he arrives. He'll start your people on the road to heaven."

"Up in the sky?"

"No, not in the sky, right here, on this planet. I know, some of what I say you can't even relate to. Well, all the changes to come won't happen in your lifetime, even during the lives of the next few generations. One thing is sure, the change will happen. Your son's the one who'll ignite the fire that cleanses the souls and minds of all those inhabiting this world. He'll even know that the Korvags and the Chroms are not the only peoples. He'll also know how to build ships that can transport your people across the big water and onto another land that has many more people than the Korvags and the Chroms together. Some of those people are decent.

Many of them, however, are worse than the Korvags. Not to worry! Your son's spirit will change them."

Bria regained the strength of her voice. "I haven't heard that any of our ancestors ever experienced anything similar to what I'm experiencing right now," she said. "I'm just hoping that I'm really not out of my mind. I've been imagining all sorts of things lately, you know. I've even tried to envision my late husband coming home and being here with me again."

"Your husband's dead, that can't be changed. He won't be coming back to you. Your daughter will."

"Will she?" Bria cried out with a sudden cheer in her voice.

"You need to be patient. You also need to be careful about who you tell what you heard here tonight. People might think you're crazy. Better not to say a word until your son returns."

"As you wish. I'm a strong woman. I can keep my mouth shut."

"I have to go now. You won't see me again." Sort of mumbling it only to himself, the man added: "From now on, as long as you live, I'll be just an invisible hologram."

Bria heard everything he said.

"A hologram... What's a hologram?"

"Oh... it's... just another word for God."

The image disappeared.

Bria spent a long time staring at the dancing flame of the candle, her brain trying to sort out details of her encounter.

Wrix did not leave. After reducing his image density, thus becoming totally invisible to organic eyes, he observed Bria struggling with her thoughts.

He could very well relate to the emotional and mental pain the woman felt.

60

'It's so much easier for me,' lamented Wrix. 'I never engage in personal relationships. Consequently, I neither suffer, nor enjoy, the effects of such interactions.'

'Obviously, it would make sense to be an organic entity if one's enjoyment and happiness outweighed the negative experiences. Naturally, the enjoyment and happiness would have to be of a highly intelligent nature perceived by one's highly evolved senses. What's the value of crude fun, especially, if it's registered only by lowly instincts? The Korvags seem to have plenty of fun. They certainly think of themselves as happy and fortunate people. Would I like to be one of them? Absolutely not...'

He decided he would not like to be one of the Chroms, either, at least not for the time being.

He felt sorry for Bria. All that pressure inside her, generated by the weight of her pain, was almost more than what her soul could withstand without getting crushed.

Wrix thought it was time to shoot a low frequency Q4 beam into the woman's brain to alter her thought process a bit. The interference immediately lessened Bria's suffering.

He watched Bria getting ready for bed.

'Still so young... and strong... and desirable,' he thought when the woman unbuttoned her long, one-piece cotton dress.

Bria pulled her arms out and then simply dropped the dress behind her onto the straw mat she was standing on. She turned to face the candlelight and for a moment she looked at her breasts. She touched one of them gently, the way her late husband used to.

"Why?" she sighed. "Why am I left alone?"

She glanced at the door as if expecting someone to enter.

'It must be extremely hard for her without her man,' Wrix thought. 'If I wasn't created without reproductive organs, I could try to soothe her aching desire. She would probably let me do that... Of course, I can be with her in her dreams ... and there she has no choice.'

Suddenly, Wrix realized he was way out of line.

Bria put on a long, white shirt and went to sleep.

Wrix left at sunrise. He beamed himself back to Mount Ulvi to see how Zuppo was doing.

'Sleeping... He'll be really hungry when he wakes up. I'd better go and collect some berries for him down on the slopes.'

While picking the fruits, Wrix felt some apprehension. 'Gods aren't supposed to do physical labor.' But then he had different thoughts: 'Come on, Wrix, would you know what Gods are or aren't supposed to do? Anyway, you could have produced this fruit synthetically.'

Back in the cave, he put the berries on top of a stone formation that looked pretty much like a table.

It was mid-morning when Zuppo finally woke up.

"Good morning to you," Wrix greeted him.

It took some time for Zuppo to remember what was going on.

"I wish you weren't invisible again," he mumbled, rubbing his eyes.

"Your breakfast's ready... Right here."

After being able to orient his eyes in the direction the voice came from, Zuppo found the fruits.

"Delicious," he commented. He devoured a large quantity in a short time.

"Do you like nuts?" Wrix's voice came from the other side of the stone table.

"Yes, I do. I also like roasted grains."

"Would you like them now or later?"

"Later. I'm full now."

"How do you feel?"

"Quite well, compared to what I went through in the last couple of days, I feel like I'm in heaven." He touched the belt over the wound in his leg. "It's not even painful anymore."

"You can remove that belt now. Your wound's healed."

"Really..." Zuppo sounded skeptical. Nevertheless, he quickly freed the area of the snakebite. "How could this happen?" he asked in amazement after peeling off the last of the leaves. A scar was all that remained.

"Next, I'd like to show you around in your new dwelling. There're a couple of other caves here as well. Follow my voice. Come this way!"

Zuppo entered another chamber.

"A pool?" he wondered loudly.

"That's right. A small underground creek feeds it. Use it anytime when you feel like swimming or cleaning yourself. The water is clean. You may even drink from it."

"I thought I was on the top of the mountain," Zuppo mumbled. "How does the water get up here?"

"It would take some time to explain. Do you really need to know?"

Zuppo shook his head. "It's not that important." He realized that he was not there to ask too many questions. Although he would have really liked to understand the reason the stone walls were emitting light inside the mountain. He had been in caves before. They were all dark.

"Very well," the voice said. "Now, go to the end of the pool. There, you have another passageway."

Zuppo found the third cave. Here, there was no need for artificial light. A huge, round hole above allowed in plenty of sunlight.

"Blue sky..." Zuppo uttered.

His surprise grew as he looked around.

"A wooden table..."

Next to the table, there was a cube-shaped straw bag covered with a thick, soft-looking sheepskin.

Zuppo's attention turned to the items on the table.

"A feather... Walnut tree extract?"

"That's right, Zuppo... and a book."

"A book," Zuppo repeated as he touched the top page with the palm of his hand. "This isn't cotton!"

"You're right, this isn't cotton. It's from wood. It is called paper. You'll learn how it's made."

Zuppo's fingertips moved to the large letters on top of the page.

"Principium," he read it slowly. His fingers moved down the page. "By Zumaniel... Prince of the Chroms..."

"Turn the page," Wrix said.

Zuppo looked at the second page which was blank. He ran his fingers on the edge of the pages.

"There're many pages there for you, Zuppo. They're all blank. You'll fill them. Provided you're really ready for the task. If not, you can simply walk out of here, go down the mountain and return to your normal life."

"Never," Zuppo responded firmly. "I'll do whatever it takes to free my people and let them live in peace."

"That's the spirit! You're the one! I have no doubt you can write this book. Once you complete the Principium, you'll take it with you to change the hearts and minds of all the people inhabiting this world."

"So, my name is now Zumaniel?"

"Zumaniel, the messiah..."

There was silence for a while.

"What am I going to write?" Zuppo asked later.

"You know very well what to write. Everything's in your mind. No one will disturb you here. You'll have peace and quite. The openings to the outside are sealed with invisible shields. No living creatures can cross the shields without burning up. Look at the lower edge of the hole over your head! Can you see that small, bright box attached to the stones?"

"I can see it."

"You'll find the same box at the main entryway. It's on the inside at your eye-level. You simply touch it once to deactivate the shield and you're free to leave the cave. When you return, touch the box again to re-create the shield."

"What if I forget to touch the box when I walk out?"

"A beeping signal will warn you when you get close to the shield."

Again, Zuppo shook his head in disbelief.

"With God, everything's possible," he muttered under his nose.

What happened next startled him even more. As his glance returned to the blank page of the book, he thought he saw a latent image of writings on it. He had a hard time believing his eyes. A full page of characters, made up of some sort of glow that was whiter than the white of the paper.

"The commands of God," he read it aloud from the top of the page.

"Go ahead Zuppo. I should now say Zumaniel. Go ahead. Dip the tip of that feather into the extract. I'll leave you alone. You'll always find food in the other chamber. You know where the drinking water is. In case you need me, just think of me. I'll never be too far away."

"Thank you, God. Thank you for everything," Zuppo said humbly.

'Poor man,' the thought zipped through the holographic circuits of Wrix's brain. 'He's acting out of fear. It's not fair. He shouldn't be subjected to feelings of inferiority just because his knowledge is limited!'

Wrix decided it was time to take solid form again.

It scared Zuppo. "Did I... say something wrong?" he asked anxiously.

"Not at all, Zumaniel, not at all, I was only wondering how I could help you to be more at ease. After all, as long as uncertainty casts a shadow over your thoughts, you can't be fully effective as a messiah. I think it'd be much better if you thought of me as a friend. Forget the notion of God! Call me friend! Here, let's shake hands!"

Somewhat hesitantly, Zuppo reached for Wrix's extended hand. When he touched thin air only, his face turned pale.

"Oh, wait!" the hologram said in a big hurry. "I'm sorry. I forget to increase my density to the necessary level. All it takes is a mental command... There we go! Now, can I see your hand again?"

At this time, Zuppo held a real, firm hand in his.

'Mind-boggling,' he thought.

"Well, I must go now," Wrix said and he disappeared.

For some time, Zuppo kept staring at the spot where Wrix stood. Later, curiously, he looked at the blank page again.

"It's still there," he moaned, referring to the glowing image on the paper. He looked at the following page. "Blank. Of course, I have to finish the first page before I know what goes on the next one."

He wetted the tip of the feather with the almost black extract and started writing: The commands of God.

While Lella was enjoying music with her headset on, Ravar made himself busy analyzing data from the planet.

'What a sophisticated technology this Wrix is,' he thought. 'A holographic image with sexual desire... Wow! Congratulations Professor Sumee! You sure have created a great program.'

He spent some time studying the frames that showed Bria's naked body. 'Not bad, not bad at all' he mused while sending a stealthy glance at Lella.

Later, he was looking at pictures of Zuppo. 'The boy is definitely a great medium. He'll make an ideal messiah.'

Ravar did some programming and increased the intensity of the bioblaster. 'The rest of the Chroms will have to be prepared to receive their Prince.'

His thoughts were interrupted. "Warning!" the metallic voice of the computer announced. "Five point five is the maximum safe level using the bioblaster to influence organic brains."

Ravar requested an explanation.

"Too much interference could make the Chroms suspicious. They would start analyzing their own actions and questioning their own thoughts. They might even start doubting that their thoughts were generated by their own brains."

"Computer, set level at five point three!"

"Level set at five point three."

Ravar got himself comfortable in his seat, closed his eyes, and let his fantasy run wild. He imagined himself lying right next to Bria.

"I'll kill you if you don't speak!" the Korvag soldier yelled at Frugo.

The old man was tied to the trunk of a tree with his feet hardly touching the ground. He had to reach down with his toes from time to time to get some relief from the pain the rope caused.

"You tell me where your leader is and you are free to go!" the Korvag repeated the offer Frugo had heard so many times during the days of interrogation.

Frugo made up his mind the day he was dragged out of Tolup: no talk with the captors as long as he is a prisoner. He had kept his pledge in spite of all the torture the Korvags subjected him to. He had been beaten several times in a variety of ways. However, being tied up and left under the scorching sun without a sip of water was worse than anything else he had to endure. His naked body was badly burned by the rays of the sun and that made the wounds inflicted by the kicks and slashes ache terribly.

The other Korvag soldier was not treating Zorel any better. The young man, also stripped of his clothes, was left hanging from a tree. Tied to a branch by his ankles, his shoulders reached down to the ground. He had a hard time finding a good position for his head. He, too, refused to answer any questions.

Frugo could see Zorel's back but he was not close enough to hear what the Korvag spoke to him.

Zorel had been whipped many times. His body was covered with long, bleeding wounds.

When Sobor arrived at the scene of the interrogation, he wanted to know whether the torture has resulted in any progress.

"Sorry, Chief, nothing so far," the soldier assigned to Frugo reported.

Sobor turned to Frugo. "Listen to me, old man! I have no pleasure in doing this to you. I'm sure you aren't enjoying yourself, either. Why don't you tell me where Zuppo is! We don't want to kill him. We just want to have a talk with him."

Frugo did not even look at the Korvag leader.

"Stubborn, very stubborn," Sobor said with a grain of admiration in his voice. He then turned back to the soldier and lowered his voice so that Frugo would not hear him: "I don't want him to die. Not yet. Take him back to the cell. Go and get the young one, too. Give them plenty of food for a change. You know what? Give them some red wine as well. Let them regain some of their strength. We have now gone too far with this rough treatment. I'll have to think of a better way to get them to talk."

Back in their cell, Frugo and Zorel collapsed onto their bunks.

The cell, one of the chambers in Sobor's house, was large enough to hold four massive bunks. The only entrance to the cell was a large, square-shaped hole in the wall, leading to another chamber. There was a gate at that opening, made from beams of wood. The beams were joined together and reinforced with ropes. The gate was simply placed against the wall on the other side to cover the hole, and it was secured to the wall with ropes. It took four of the Korvag men to lift the gate and move it to the side when someone had to pass through the hole.

Two soldiers, stationed in the other chamber, guarded the gate day and night.

A torch on the wall, right above the gate, lit up the chamber.

"Food for you!" one of the soldiers shouted as he tossed in bread and roasted duck meat between the beams. "You'll have a feast tonight. Come and eat while it's still warm. I think it's quite a waste to feed you duck meat but that's what the chief sent."

Zorel, limping from his bunk, picked up the food.

"Wait!" shouted the guard. "Take this, too!"

It was the usual jug the guard handed in.

"Today, it is wine," the guard said. "The wells must have dried out."

Zorel gave half of the food to Frugo.

They ate quietly, both of them sitting on the old man's bunk. As usual, they shared the content of the jug.

"I wish it was water," Frugo broke the silence. "This strong wine will knock our brains out in no time."

"We have no choice, we must drink it," Zorel replied. "The sun dried us out, we need the liquid. It'll also keep us warmer during the night."

As Frugo predicted, the wine got them off their feet. Zorel returned to his bunk and they both stretched out.

They did not get the chance to fall asleep.

A female's shrieking made them sit up and look through the opening. They saw two men carrying someone to the gate.

"Zea," Zorel said in surprise when he recognized Zuppo's sister.

The girl was trying to kick and fight but was helpless against the men.

"What's going on?" Frugo asked, rubbing his eyes.

The guards opened the gate and then closed it after they pushed Zea inside.   The girl had also been stripped of her clothing. She cried and tried to cover the area of her groin with one hand and her breasts with the other. She appeared to be in a horrible emotional state.

"Go on!" the Korvags barked and laughed on the other side of the gate. "Go and get them! Make love to the old one first! He's ready for you! Go on!"

Zea fell on her knees and then she collapsed onto the stony floor. Her entire body was twitching as she sobbed uncontrollably.

"Zea," Zorel called her name.

The young man managed to get on his feet but then he fell down after the first step. He crawled to the girl and tried to lift her head.

Zea looked up when she felt the hand gently touching her face.

"Oh, Zorel," she cried through her tears.

Zorel raised himself into a kneeling position and helped Zea to do the same.

They looked into each other's eyes painfully. A moment later, Zea broke down again and collapsed into Zorel's arms.

"Go ahead! Do it, man! Now or never!" the barbaric yelling came from the other chamber. "She's all yours now! Get her!"

"Help me, Frugo," Zorel called the old man.

Together they lifted Zea and laid her on Frugo's bunk.

"That's right!" one guard shouted. "The old man goes first!"

"We'll share my bunk tonight," Zorel proposed to Frugo. "It's wide enough for the two of us."

Just as Frugo nodded, a harsh cry overpowered the guards' voices.

"What are you doing to my darling?"

It was Sobor standing at the gate.

"Open it!" he yelled at the guards.

Once inside, the Korvag chief marched to Frugo's bunk and grabbed one of Zea's arms. "You'll die for this!" he roared, directing his words at Frugo.

Zea shrieked again as loudly as she could.

Sobor could not care less. He pulled the girl off the bunk. When she fell to the floor, he lifted her roughly into his arms and moved her out of the chamber.

"Secure the gate!" Sobor gave the order to his guards.

Zorel and Frugo soon fell asleep and slept through the night. When they woke up in the morning, they saw Sobor standing there in the middle of their prison chamber.

"I hope you slept well," the Korvag chief said, "because this might have been your last night alive. What you did here in the evening is definitely inexcusable. First, you were unable to control yourself when you tasted the wine I sent. You drank it all and lost your mind. I had my soldiers bring in the Chrom girl to care for you, to keep you alive." Sobor paused and when he continued talking, he roared. "You bastards raped the poor girl! Both of you, you animals." He lowered his voice as he went on. "I should have slaughtered both of you immediately after my soldiers reported what you had done... Well, I had a better idea. I'll hand you over to your people and inform them of your hideous crime. I'll let you be judged by

your own. They will surely stone you two to death." Sobor was looking at the faces of the two Chroms to see what effect his words had. "Of course," he continued, "you can avoid such humiliating death if you simply tell me where Zuppo is. You just tell me where I can find him, we'll bring him here, we'll have a talk with him... and then all three of you can safely return home... Otherwise, you know what awaits you. So, what do you say?"

He stepped to Frugo and grabbed one of the old man's ears with his fingers.

Zorel raised his hand to draw Sobor's attention."Well, I'm listening," Sobor said, letting go of Frugo's ear.

"We did not commit that crime last night," Zorel started. "Your accusation is nonsense. Our people would never believe you. We are not Korvags."

"Is that all you have to say?!" Sobor shouted.

"That is all I have to say," Zorel said.

The Korvag turned to Frugo again. "How about you?!"

"I have nothing to tell you," replied the old man.

Sobor was quiet for a while. Then he yelled again: "I'm done with you two. What you deserve is coming to you… Unless you change your mind," he added as he stormed out.

In the absence of Zuppo and Zorel, Morva took over as leader.

The Sevens, now down to five, continued their meetings at dawn.

With the exception of the very young and the very old, the people of Tolup learned how to write and read. They started recording their songs and their stories. Most of their writings were about their desire to live freely and without fear.

Tiho, the old man who still refused to have an animal to pull his cart, wrote his own little song on the back of his shirt:

'In the morning when the stars go to sleep,
The sun wakes up, a new day begins,
I pull my cart and go to the field,
Hope that one day we can all live in peace.'

A Korvag soldier, one of those sent to Tolup to collect the crop when harvest was over, noticed the writing on Tiho's back and immediately reported it to his commander.

"Just what's that strange marking on your shirt supposed to be?" the commander asked Tiho.

"It's my song," replied the old man.

"Your song... Can I hear it?"

The collection took place in the village-square. There were many other people in line behind Tiho, bringing their goods mainly in mule-drawn carts.

Tiho realized it had been a mistake to let the Korvags see his song in writing. He hesitantly turned around and looked at the man next in line that happened to be Toma, one of the Sevens.

Toma shrugged. "Go ahead, Tiho, sing it."

The old man took off his shirt and laid it out on a sack full of grain, eliminating the creases with the palm of his hand so that every character was clearly visible.

While singing, he moved his fingertip along the lines.

The Korvag commander wrinkled his forehead as he got lost in his thoughts for some time after Tiho finished his performance. Eventually, he looked at the Chroms waiting there in the long line and shouted: "Does anyone here or anyone you know has markings similar to this?" He raised Tiho's shirt above his head, stretching it with his hands so the people could see the song written on it.

There came no answer.

The commander nodded after some more pondering, and then, quite unexpectedly, he began yelling at Tiho: "You shriveled sun of a gun! You think you can fool me and not get killed? You are dead wrong!" He swung his ax and put a sizable hole into Tiho's skull.

With blood erupting from his ear, the old man fell to the ground and did not move anymore.

After the commander stuffed Tiho's shirt into a pocket on his horse's saddle, he turned to the people in line: "Does anyone else want to sing?"

The crowd was silent.

Toma clenched his fists behind his back and forced himself to stay calm. He knew his people would have no chance fighting the Korvag soldiers. It would be senseless losing even more lives.

"Remove this clown from the square!" the commander ordered Toma. "Then hurry up with those sacks! We don't have all day."

As one Korvag rode out of the village with his cargo hanging from the back of his horse, another one came to replace him. After the last Chrom turned in his goods, the Korvags left Tolup.

The entire village mourned Tiho's death.

"He was a gentle man," Morva remembered him during the meeting of the Sevens the following morning. "He's never harmed anyone, not even a fly. He certainly didn't deserve to die such horrible death. We'll remember him as long as we live."

"How long can we go on living like this?" Toma asked with suppressed anger in his voice. "Can't we do something?"

"Let's wait until Zuppo returns," suggested Morva.

"What if he never comes back?" Grog argued. "He could be dead by now. Can he survive out there all by himself?"

"Perhaps we should go look for him," commented Bodor.

"I believe he's all right," Morva said. "I saw him in my dreams. He wasn't alone ... I think he was with God. He'll be back when the time arrives. He'll be back to lead us."

"I hope he comes soon. Winter is harsh on the mountain."

"We don't know when he comes. We have to wait patiently. He'll be all right. Winter can't harm him. He's safe up there, God takes care of him."

Later, they talked about the Registry that Zorel's brother, Miran invented.

"It's very clever, indeed," Morva said. "Now we can record the passing of each cycle from the beginning of summer to the beginning of the next one. We can also keep score of the days within the cycles. Miran makes notation of every significant event as the days go by. Yesterday, for instance, he marked down that Tiho's life ended. Today, he'll record the birth of his own son."

"Let's just hope the Korvags never find out about the Registry."

"Miran keeps it hidden. It's in a safe place."

When they were ready to return to the village, Morva said a short prayer asking God to protect their people held captive by their enemies. He finished the prayer the same way they usually did: "...and we ask God to deliver us from the claws of the Korvags."

A few days later, Miran was invited to attend the meeting of the Sevens. When the sun came up, it was his turn to speak.

"As you can see, I have two books with me," Miran began his presentation. "The bigger one is the Registry. I have already filled the first two pages. I'd like to read my notes to you so that you can tell me what else from our past should be recorded."

After he finished reading, and the leaders provided some useful advice, Miran opened the second book.

"I found it necessary to work out a system of values," he continued. "I believe that one day we'll be strong enough to refuse the demands of the Korvags. From that point on, we should insist that they repay everything they looted from us. I established the basic unit as one bucket of corn. If none of you object, I'll call the worth of this unit a Tolur, naming it after our village. So far, our people have exchanged goods without paying much attention to real value. All of us should start using the Tolur. This would help us later when we're in the position to hold the Korvags accountable. Of course, everything will have a price. For example, as long as we equate five apples to a bucket of corn, ten apples would be worth two Tolurs. With some help from the elders, I can create a price list to show the value of all goods. We may want to use the yellow metal we found at the foothills as the actual medium of exchange. After we melt it from the ore, we can easily mold it into small disks. The size of a thumbnail, not thicker then pig skin could be the equivalent of one Tolur. I've estimated that after we extract all the metal from the ore deposits we've found we can have enough to cover the value of all commodities in our domain. Every household then receives their fair share of Tolurs. I'll carefully calculate the debt of the Korvag people going back three generations and keep the accounting in this book. When they start repaying what they owe us, we can deduct the value of goods we receive. Once they paid all their debts, their balance will go to zero."

"They should pay us more than what they owe," Grog suggested.

"I agree," Miran said, "perhaps a tenth more."

"At least that much," Morva raised his voice. "Then perhaps we can overlook all the pain and suffering they caused us."

"Not to mention the lives they took!"

"Very well," Miran replied. "I'll add ten percent to their total debt. We might even consider an additional one tenth of the ten percent on the outstanding balance once every summer. This would encourage them to eliminate their debt as quickly as possible."

"There's one thing I don't understand," Toma said. "Would the Korvags pay us in goods or in Tolurs?"

"It's quite obvious that they couldn't repay everything in goods," Miran answered. "Even if they turned over to us everything they currently possess, it would cover only a fraction of what they owe us."

"There is no way they could pay us in Tolurs," commented Grog. "They don't have any of the yellow metal."

"Not yet," responded Miran. "First, we'd have to show them how to search for the ore. I'm sure some can be found in their mountains as well. It might take a lot of hard work. That's exactly what we'd like to see the Korvags do for a change, a lot of hard work. Am I wrong?"

They all agreed with Miran.

Morva said his usual prayer and then he adjourned the meeting until the following morning.

The lukewarm days of autumn witnessed a lot of diligent activity in Tolup.

The Tolurs were cast and distributed among the inhabitants.

Miran came up with a reasonable price for everything. A hen, for instance, could be bought for four Tolurs. A mule's worth was set at one hundred Tolurs.

Piska, one of the elders, made a significant discovery. He found that a salty substance he unearthed in a nearby dried out lake was highly flammable. This gave him the idea of building small rockets. He simply dissolved the salt in water and then he boiled the solution until most of the water evaporated. He mixed the concentrate with dry sawdust and filled the mixture into cone-shaped birch-bark reinforced with yarn on the outside. After several days on the sun, the device completely dried out. Ignited at the opening on one end, the explosively burning core sent the rocket way above the treetops.

The leaders believed that Piska's discovery would play an important role in the future of the Chrom people.

Autumn gradually gave way to the cool, rainy days of winter.

The Sevens got permission from the inhabitants to set up their headquarters in the late Tiho's house. They now held several meetings during every day, consulting with anyone who had new ideas about how to change the course of Chrom history.

The sky was overcast most of the time. On rear occasions, when the winds blew away the clouds and the sun dried up the fog, people looked at the snow capped peak of Mount Ulvi, wondering whether Zuppo was still alive.

Finally, spring arrived. The warm winds cleared the sky and melted the snow. The Chroms got busy getting the soil ready for the next crop.

One day in the middle of spring, when everything was in full bloom, a couple of kids were playing just outside the village wall. They were chasing butterflies, running around barefooted in the fresh grass. Suddenly, one of them stopped and looked in the direction of Mount Ulvi.

"What is it?" asked the other one.

"Look! Someone's coming!"

In the next moment, both of them were dashing back to the village. After climbing over the wall, they started screaming at the top of their lungs: "Someone's coming! A stranger's coming!"

People who were not out in the field, gathered at the gate.

Bria was among the first ones to arrive. When she saw the man approaching, she immediately recognized him.

"Zuppo, my son!" she shouted happily. With tears in her eyes, her arms open, she started running.

Zuppo ran, too. He wrapped his arms around his mother, lifted her up, and turned around with her several times.

"It's so good to see you, Mam."

Bria was laughing and crying at the same time. "I can't even kiss your face," she said referring to her son's overgrown beard. "My God, you've changed so much. If you weren't my son, I'd have a hard time recognizing you. Look at your hair! It has all turned gray. Your face looks hardened. You must've had a rough time since you left. What's that under your belt?"

"This is the book of God, Mam," Zuppo answered in a calm voice.

"Can I see it?"

Zuppo pulled out the bulky book and handed it to his mother.

Bria looked at the writing on the cover and read slowly: "Principium... by Zumaniel.... Zumaniel..." As she said that, she hesitantly pointed her finger at her son.

"Yes, Mam, that's me. That's how God wants me to be called from now on."

The woman curiously tested the pages of the book between her fingers. "I've never seen anything like this."

"It's paper, Mam. Paper is made from wood. I have the recipe for it."

"Come! The people have been anxiously waiting for you."

"Blessings to you all," Zuppo greeted the onlookers as he walked through the gate, "blessings to you all."

"He grew old... He looks so powerful... He's definitely our messiah," people whispered as they followed Zuppo back to the village.

"What has changed since I left the village?" Zuppo asked his mother.

"Oh, Zuppo... Zumaniel... Many things have happened while you were gone. I'll tell you everything when we get home... You must be hungry."

"I've missed your cooking, Mam."

Once inside their house, leaving the crowd out in the street, Bria put her nicest cloth on the table.

"I'll bring some fruits from the pantry."

Zuppo stopped her. "No fruits, Mam! No fruits, please. I'd rather have some of your porridge."

Bria glanced at the stone fireplace under the chimney. "Luckily, it's still burning. I'll warm up the porridge for you. I cooked it this morning."

"Have some milk and butter?"

"They are all fresh, my son. I'll have them ready in no time."

"While you're doing that, I'll bring water from the well and wash the dust of the road off my body. Where do I find the wood basin?"

"Go get the water. I'll bring the basin... If you want, I'll come and wash your back. Just as when you were a kid... Oh, Zuppo! Am I glad you're home! Come and let me hug you again! I've missed you so much."

Zuppo embraced his mother and held her in his arms.

This was one of those days when the Sevens held no meetings as they, too, were plowing their parcels. They first heard about Zuppo's return upon their arrival back in the village after sunset.

Morva wanted to meet their leader right away.

"Supper can wait," he told his wife and hurried over to Zuppo's house.

Bria and Zuppo were about to start their evening meal when Morva arrived.

"Just in time if you care to join us," Bria welcomed him. "You must be hungry, too."

"Thanks Bria. You know I'm always hungry but seldom have time to eat. I just came to see my dear friend, our respected leader." He turned to Zuppo to shake his hand. With his arms extended, he rested his hands on Zuppo's shoulders. "Let me take a good look at you! I'm happy to see you're alive and well. Welcome back. We've been worried about you. We've even thought the Korvags might have captured you. Last time they were here, they searched all over for you. I'm afraid they want your head. Therefore we created a secret bunker for you."

Morva paused and then he went on. "We haven't seen any Korvags since they came for their loot last autumn. As you know, they very seldom come over the swamp during the winter... but one never knows. Your bunker is under Frugo's house. There, you can be safe day and night."

"Morva," Zuppo interrupted. "I didn't come back to hide. It's not necessary for me to do that. I'm not afraid of the Korvags anymore. They can do no more harm to me. My faith in God is very strong. No brute force can overcome God's power. Give me some time and you'll understand everything I say." He reached for his book and handed it to Morva. "Take a look at this."

Seeing the thin pages of paper, Morva got curious. "This isn't leather, not canvas, either."

"It's paper, Morva. Paper, made of wood," Zuppo explained.

"The Commands of God," Morva read when he got to the second page. "Command number one. Never take the life of a warm-blooded creature! Killing another human is the greatest sin of all. There can be no reason, nor justification for such horrible act. Whoever commits the act of murder destroys his or her inner self, the soul, at the same time. Life on this planet is nothing more than preparation for a wonderful experience in another realm. Dead souls are barred from that realm. Dead souls are dead forever... Killing a warm-blooded animal is a sin of second degree. When drought, flood or other natural catastrophe destroys your crop, you may kill such animal to feed on its flesh. Don't get used to it though, because it damages your soul, confuses your mind, makes your will ill tempered, and weakens your body. Cleanse your body of any residue as soon as you harvest your next crop. Eat fish which is in abundance in the rivers and lakes, and especially in the ocean."

Morva looked up from the book. "What is ocean?"

"Very big water," Zuppo answered. "You'll see it one day when we set sail for our new homeland. It's all explained in this book."

"It's a pity we have only one book."

"We'll make replicas. The older folks who no longer work in the field can write those replicas. First, however, we have to produce paper."

"How can we do that?"

"The recipe is on one of these pages."

Morva nodded and continued reading. "You may take the skin of animals that die naturally or get killed by other animals. However, best to avoid letting your own skin get in contact with the skin of a dead animal. Surround your body with material that had its roots in the soil."

"Wait!" Bria said. "What's that commotion outside? Can you two hear it? Or is it just my imagination?"

"You're right," replied Morva. "Something must be going on out there."

"Korvags, Korvags!" shouted someone in the street.

Bria sprang from the table. "I'll extinguish the candles!"

Zuppo stopped her by raising his hand. "Mam, don't! It's not necessary. They won't do any harm to us."

"Open the door!" someone barked outside.

"It's open," Zuppo shouted back.

Bria and Morva could not believe their eyes when they saw the latch move without anyone touching it.

Three Korvag soldiers entered the house.

"I'm Zebrio, one of Sobor's commanders," the tallest of the three introduced himself. "You, the one with the long beard, you must be Zuppo!"

"Yes, Zebrio, I'm Zuppo... Actually, I'm Zumaniel from now on."

"I don't really care what your name is. Sobor sent us to arrest you and take you to him. Don't try to escape! My men surrounded the house. You can't run. The entire village is under my army's control."

"Listen to me, Zebrio!" Zuppo replied. "Take your army and go in peace! Go back to your domain and tell Sobor to let the Chrom prisoners return to Tolup!"

Morva was shocked. No one ever talked to a Korvag like this before.

"I mean you'd better surrender!" the commander said after a brief pause.

'A pleading Korvag...' Morva wondered. 'Am I dreaming?'

"Go home, Zebrio! Take all your men and go in peace!" Zuppo said.

The commander stepped forward, threw a bundle of ropes on the table and looked at his men. "Tie him!" he gave the order.

"Stop right there!" thundered Zuppo.

A thin beam of bluish light, originating from where Zuppo stood, suddenly paralyzed the three Korvags. For a few moments, they were unable to move. Only when the beam disappeared were they able to resume their motion.

"Zebrio, do you want to get frozen again and remain that way forever?" Zuppo asked.

The Korvags were all confused. Slowly, they were backing towards the door.

"Take your rope!" Zuppo yelled at the commander.

"May I?" Zebrio asked in a shaky voice.

"Take it!" Zuppo roared.

"I will... I will... Don't freeze me again... I just came here because Sobor sent me." Stuttering and fearful, Zebrio inched his way forward until he could reach for the rope. "I'll just take this... and leave... I never meant any harm... Forgive me."

"How does Sobor know that I've returned home?"

"He didn't know... He just decided to send us to see if you were here."

"Tell him not to send any more Korvags unannounced!"

"I got it."

The other two were already outside.

Zebrio, as soon as he left the house, regained the strength of his voice. "Retreat, retreat!" he yelled at his soldiers. "Everyone, get out of Tolup, back to Dudvo!"

"We'll sleep undisturbed," Zuppo said after the outside noise distanced away. "Go home, Morva. Go home and rest. Tomorrow, I'll go over the swamp. You may join me if you wish."

"I will," Morva rushed to answer.

"Let's meet at the gate shortly after sunrise."

"I'll be there waiting for you. Good night."

"I'll put the porridge back on the stove," Bria said after Morva left.

When she returned to the table with the steaming cereal, she found her son burying his face in his hands, praying quietly. She waited until Zuppo finished the prayer. They ate without saying a word.

Later, Bria asked: "Son, how did you do it?"

Zuppo did not answer right away. "I didn't do it, Mam", he said then. "I only wished it would happen... It was God's doing."

Wrix sat in the corner of the prison chamber. He watched the Korvag guards opening the gate and allowing half a dozen soldiers to enter.

Two of the soldiers tied Frugo to his bunk while the other four began to beat Zorel.

'Senseless!' Wrix thought, 'absolutely senseless!'

He had seen the same happening over and over again ever since Frugo and Zorel were captured: interrogation almost every night.

'The only human gesture the Korvags have demonstrated is that they gave warm clothing to the prisoners during the winter. Otherwise, they are brutal.'

The soldiers punched Zorel in the stomach several times.

Shaking his invisible head, Wrix continued his pondering: 'If I were a free agent, I sure wouldn't tolerate the Korvags' primitive behavior. I would numb all of them and leave them paralyzed for a long time... What logic does Ravar use when he programs my actions? If he wanted to, we could eradicate these monsters in the blink of an eye.'

As usual, Zorel tried to fight back but he was no match for the soldiers.

Frugo, who had been treated less severely, could not keep his mouth shut any longer. "Animals!" he shouted. "That's what you all are, animals!"

"Shut up!" one of the Korvags yelled at Frugo.

"I won't shut up!" the old man shouted back. "What you're doing is outrageous! Zorel is a young man. He's never harmed any of you. Why are you torturing him?"

"He should speak!"

"He has nothing to say. He doesn't have answers to your questions."

"Shut up!" another soldier barked.

"You shut up!" Zorel snapped back, breaking his vow never to communicate a single word to his captors.

"Wow! It seems like we have a rebellion on our hands," another one of the soldiers said guffawing. "Let's see who shuts up first." First he kicked Zorel in the area of his groin and then he punched his throat.

Zorel collapsed and passed out.

"Animals," Frugo shouted again. "You're lowly animals!"

"That's enough, old man!" The guffawing Korvag's face turned red from anger as he stepped to Frugo's bunk. "Let's just make sure that you keep quiet from now on." He wrapped a rope tight around the old man's neck.

The soldiers were all laughing when they left the chamber.

By the time Zorel regained consciousness, Frugo was already dead.

'Senseless,' Wrix confirmed his previous comment, 'absolutely senseless. Killing another innocent man... Is there no real God assigned to this planet?'

Wrix watched Zorel trying desperately to revive Frugo.

'It's too late,' he thought sympathetically.

"Why?" Zorel cried out loud. "Why did I have to break my silence? Why?" He lowered his head and rested it on the old man's chest. "Good old Frugo, you could still be alive... Your death is my fault."

'That's right,' Wrix agreed. 'He could still be alive... You should've remained strong and defiant, just as you had been... You sure would've, had you known about those Korvags I paralyzed over at Zuppo's house. They are just arriving back here in Dudvo... Once they tell the people of their strange experience, conditions in this prison ought to change... Right, if you could have held out at least for another night.'

It was time for Wrix to leave the prison.

'I'm curious to see what kind of reception Zebrio will have,' he wondered while beaming over to Sobor's house where the returning commander had to report.

The big chief of the Korvags already had supper, and he also had plenty of red wine to wash it down. He was in a relatively good mood, relaxing in his favorite chair in the front chamber of his house. Normally, at this time, he would be in bed with one of his concubines, making love either passionately or forcefully, depending on whether he desired a Korvag or a Chrom girl. This day was exceptional.

"Zebrio should've returned by now," Sobor grumbled to himself.

Finally, a guard stormed in to tell Sobor that Zebrio was waiting outside.

"Let him in!" Sobor instructed the guard.

"Chief, I'm sorry." This was the first thing Zebrio said after he entered. "I must tell you that..."

Sobor promptly interrupted him: "Don't tell me you came back without the Chrom!"

Zebrio hung his head, sighed, and opened his arms.

"What's that supposed to mean?" Sobor shouted while standing up. While walking around the massive stone table, he kept looking at his commander, waiting for the answer.

"I don't even know how to tell you what exactly happened," Zebrio began his explanation. "I doubt you'll believe me... Now that I look back, it seems so unreal... Nevertheless, it did happen."

Sobor lost his patience. "Go on! Spit it out!" he roared. "Speak, man! Tell me what happened!"

"Chief, the Chrom leader has magical powers... I was ready to capture him... but he made it impossible."

Sobor stopped right in front of Zebrio and looked at him with narrowed eyes. He lowered his voice. "Go on, I'm listening."

"Chief, we surrounded the house as you ordered. The Chrom wouldn't have had the slightest chance to escape... I went inside with a couple of my best soldiers... The Chrom was having supper with a woman and another man... Believe it or not, he didn't seem to be concerned by our presence. His eyes didn't even blink when he saw us enter. That already made me suspicious... I told him you sent us to arrest him and bring him here... He just stood up and told us to leave in peace. He even said to tell you to release all Chroms and let them go home."

As he stopped to take a deep breath, Sobor took over. "So far I don't like it!" he thundered. "You sound just like our late coward leader, Nezir... So what happened?"

"Chief, you'll think I'm a liar... I swear, what I'm about to tell you is as true as the fact that I'm standing right here in front of you... I threw my ropes on the table and I stepped forward to grab him... He stood up and shouted 'stop right there'. He raised his hand, and I saw some kind of rays coming right through me. It completely paralyzed me. The rays I mean... I couldn't move. My brain was working, I was able to see and think... but I couldn't get my body to make any movements... When he lowered his hand, the rays were gone, and I regained my mobility... He said if I didn't want to get paralyzed again, I should leave immediately. So, I did."

"You idiot," Sobor exploded. "You are a feeble-minded, stupid idiot!" He leaned closer to Zebrio and sniffed at him. "You must have shit in your pants. I should kill you right here for feeding this fairy tale to me. Do you think I buy this sort of non-sense? You know me better than that." He paused for a moment and then he called in the guards. "Remove this coward from my chamber! He's a traitor. Take him to the prison and lock him up with the Chroms!"

"Chief," Zebrio tried to protest, "I'm telling you the truth. Ask Sirop and Krovi, they were with me when this happened."

With a sudden movement of his hand, Sobor knocked Zebrio to the floor. He pulled his ax from its holder on his waist and swooped down with the weapon. The sharp stone broke through Zebrio's skull as if it was a loaf of bread.

"I hate cowards," Sobor said calmly after he pulled his ax out of Zebrio's head. "Dump him outside the village!" he instructed the guards. "Let the wolves chew the flesh off his bones!"

"Do you want to see Sirop and Krovi?" one of the guards asked. "They're waiting outside."

"Tie them and take them to the prison!"

"As you wish," replied the guard.

Wrix raised his holographic eyebrows. Sitting there invisibly on Sobor's stone table, he knew his dislike for the Korvag chief has just increased at least by another notch.

The first rays of the rising sun lit up the dewdrops on the grass.

Zuppo and Morva arrived at the gate almost at the same time.

"I can't go with you," Morva told Zuppo after they exchanged greetings. "I had a dream last night warning me of grave danger. I believe it's unwise to go over to the Korvags today. You shouldn't go, either."

Zuppo listened carefully before looking toward the village-square. Someone was coming.

"It's Miran, Zorel's brother," Zuppo said.

Morva was surprised. "What's he doing here this early?"

"I also had a dream, Morva... After that dream, I suspected you'd change your mind. So, on my way here, I stopped at Miran's house and asked him to get ready. I told him to bring all his accounting so that we can hand our bill over to the Korvags. Indeed, Miran's the right person to come with me. You're needed here in the village."

"With all due respect," Morva reacted, "I think you should stay here, too!"

"I must go. I have to take God's message to the Korvag people. They need God just as much as we do. They don't even know how miserable they are without God."

"You're risking your life."

"That's possible, still, I must go. I'm the only one who can deliver the message." Zuppo opened the rolls of cotton sheets he had in his hand. "I've reproduced the most important parts of the Principium. As you can see, I'm not taking the entire book with me at this time. I left that with one of the elders. They'll have all the pages copied onto canvas. We will need a replica we can hide in a secret place. The elders are also studying the method of producing paper. Soon, we should have copies of the Principium written on paper."

Miran joined them. "Good morning to you both," he greeted them and then he turned to Morva. "I didn't know you were also coming."

"He's not," Zuppo replied. "He's on his way back to the village."

"I'd like to ask you once more to change your mind," Morva told Zuppo. "It's very dangerous for you to go to the Korvags."

"Even if I wanted to change my mind, it's too late now. I've sent a messenger to let the Korvag guards on the bridge know of my arrival. The news must have reached Sobor by now and he's expecting me." Zuppo turned to Miran. "Morva is concerned about the dangers of this mission. You may also stay behind if you want to."

"My mind is made up," answered Miran. "I'm going with you."

The two said good bye to Morva and walked toward the swamp. When they arrived at the edge of the water, a few Chrom youngsters greeted them.

"We've just finished building a raft," one of the young men told Zuppo. "No need to get your feet muddy. You simply stand on this float and push yourself with this long stick. You should be at the bridge in no time."

"Well, let's give it a try then," Zuppo said after praising the youngsters.

"I'll be the pusher," Miran volunteered. He handed his records to Zuppo and grabbed the stick. "Let's go then!"

The raft began to move as Miran leaned against the stick.

Zuppo, dressed in a long, white robe, with rolls of canvas under one of his arms, stood at the front.

Miran guided the float as if he had done nothing else in all his life. Even in the passageway the Korvags cut in the reeds he had no difficulty navigating.

When they reached the bridge, Miran secured the raft with a rope.

"Quite a reception," Zuppo commented, referring to the number of onlookers, mainly children, who gathered on the shore.

When Zuppo and Miran stepped off the bridge, a Korvag soldier greeted them. "The big chief of the Korvags is expecting you," he said in an indifferent voice. "I'll lead you to him."

The curious crowd formed a wall on both sides along the way. They were people of all ages, most of them young and old. They kept telling each other what their thoughts were about the Chrom in the white robe. Often, the comments were loud enough for Zuppo and Miran to hear them.

"He sure looks like some kind of a messenger," an old woman said.

"I bet he could paralyze all of us," whispered a young kid.

"Will he make Sobor disappear?" a girl asked her mother.

The soldiers who were sent to keep the crowd in check shouted 'quiet', or 'shut up' from time to time but they lacked the usual firmness.

"My father was there when it happened," a child told his friend. "He was just outside the house, so he saw everything through the doorway. This bearded creature just raised his hand, and Zebrio was frozen like a rock. Yeah, my father is still scared. I'm scared, too. Just look at him!"

The sun was already up high in the sky when finally the houses of Dudvo appeared. The first storm of the season was gathering strength on the horizon; angry, dark clouds grew swirling towards the sun. In spite of the approaching storm, the wind was calm. Even the thunders sounded surreal, as if someone was rolling huge barrels on a wooden roof.

"The chief is inside," a guard took over at Sobor's house. "Come after me!"

Zuppo and Miran followed the guard with a number of soldiers behind them.

Sobor was sitting in his chair, eating berries from a bowl.

"Good day to you, chief of the Korvags," Zuppo greeted him.

Sobor nodded and put another piece of the fruit in his mouth. Soon, he pushed the bowl away with the back of his hand and took a better look at the two Chroms. He scratched his forehead briefly as he leaned back in his chair.

"Good day," he said, still chewing. After swallowing what he had in his mouth, he raised his voice. "You, in that white..." He forced a laugh. "What is it, underwear?"

The soldiers laughed, too.

"Anyhow," Sobor continued, "you must be that... that big Chrom leader... Well, your beard looks impressive, no doubt... Anyhow, I'm glad you obeyed my order and came."

"I obeyed no order of yours," Zuppo answered calmly. "I came on my own."

Sobor suddenly stood up. "It won't mean any difference," he said, "especially after I killed you."

The Korvag chief was a frightening sight, tall and massive like a rock. He wore a sleeveless shirt, a short and a vest. The short was made of pig skin while the vest was of another type of leather. Leather patches protected his soles, secured to his feet by strings.

Zuppo did not seem effected by the threat. His voice did not change, either. "I came because God wanted me to come. I came to bring good news. I came to propose peace between our peoples, peace that'll make life happier and more prosperous for all of us."

"Shut your big mouth!" Sobor shouted. "I'm not interested in your happiness and your prosperity. I'm the leader of the Korvag people... I'm here to make sure that the Korvags live in happiness and prosperity... Every Chrom bastard should understand that clearly! I'm not saying including you because you won't be included... You're hereby sentenced to death! I don't even hold your gibberish talking against you. Your crime is organizing and leading that raid against my people. You kidnapped a Korvag chief! It doesn't matter that he was a coward and had to be killed for his cowardice. What matters is that you dared to disturb us. You killed a good number of the Korvag guards. You, a coward yourself, sneaked in here under the cover of darkness, in the middle of the night when most decent people are asleep. You stole our possessions."

"Your possessions," Zuppo interrupted him. "What gives you the right to kidnap our sisters and keep them as your slaves? You use them against their will. You rape them. You, the giant and powerful Korvag rape innocent, fragile, defenseless girls. Let me tell you, Sobor, you're the greatest coward of all."

Sobor's face turned red. He stepped away from the table and closer to Zuppo. His lips tightened and his nostrils widened as rage took him over.

"I should've killed you the moment you stepped in here," he barked. "I won't delay your execution any longer."

Sobor reached for his ax, grabbed its handle, and raised his hand, ready to swoop down and split Zuppo's head open. Just as he began to move his hand forward, a hardly visible beam of light, emanating approximately from under one of Zuppo's arms, ran through Sobor's forehead. His motion froze instantaneously.

Zuppo made a few steps backing away from the Korvag leader's paralyzed body. As soon as he was out of reach, the beam disappeared and Sobor could finish his motion, cutting into his own leg just above the knee.

Like a wounded animal, Sobor howled. He dropped the ax and covered the deep cut with his hand, trying to stop the profuse bleeding.

"Your shirt!" he shouted at one of his guards. "Give me your shirt!"

The guard hesitated. He was bewildered by the unexpected turn of event.

"I said give me your shirt!" Sobor roared.

"Oh, yes," the guard reacted at last. "Here it is, chief... Do you want me to tear it in half for you?"

Zuppo stepped closer to the Korvag chief. "That shirt won't be needed," he said and briefly placed his hand over Sobor's wound.

When Zuppo removed his hand, the wound was completely healed, leaving only a hardly visible scar. All blood from the Korvag's leg disappeared as if nothing had happened.

"No pain," mumbled Sobor as he stared at the scar. He then looked at Zuppo with confusion lurking in his eyes.

"This incident shouldn't have happened," Zuppo said quietly. "As you can see, God doesn't want you to suffer. God wants all of us to live in peace... You've been a very wicked man, Sobor, but God can forgive you if you change your ways."

"What do you want?" the Korvag chief said abruptly. He turned around and walked back to his chair.

"You know very well what I want more than anything else," replied Zuppo.

"Granted," Sobor answered angrily. "The girls are free to go back to Tolup. So is Zorel... The old man's dead."

"There's another thing," Zuppo said after a pause. "You've been robbing us of our goods for a long time. My fellow Tolupian, Miran, has all the records here with him. He'll tell you how much you owe us."

"What?" Sobor sprang up from his chair and cast a furious glance at Zuppo.

"There's no need to react like that," Zuppo continued calmly. "As long as you and your people follow God's commands, you won't starve or suffer any other ways."

"What commands?" Sobor interrupted again.

"I'll read important parts of the Principium to your people." Zuppo held out his hand with the writings in it. "These are excerpts from God's commands. Eventually, your people will learn to read and will have their own books of the Principium. Some of the Chroms are ready to come and be your teachers."

"That's good," one of the guards commented.

"Shut your mouth!" Sobor yelled at the guard. "I decide what's good and what's not." The chief paused, pressing his fingertips against his forehead. He turned back to Zuppo.

"Your teachers are not wanted in our land. I'll send some of my people to your village... You can show your magic to them. They'll share it with us."

"It's all the same to us," Zuppo answered. "Now, back to your debt... Miran, read those numbers to the Korvag chief."

While Miran presented the statistics, Sobor's blood pressure kept rising.

"Enough!" the chief finally exploded, interrupting Miran. Rubbing his cheek, he continued. "This must be some kind of a dream," he said in a lowered voice. "Soon, I'll wake up and you... both of you... will be gone... Trong!" he shouted at the guard standing closest to the entryway. "Look outside! See if the sun's up in the sky!"

"It is broad daylight, chief," the guard reported after he stepped back inside.

Sobor sighed, something he very rarely did.

"Chief of the Korvags," Zuppo took over again. "I realize it would take generations before you can repay us in the same goods you've stolen from us. Therefore, I have a proposition."

"What proposition?" Sobor snapped back. However, he sounded more conciliatory than anytime before.

"On my way here from the swamp, I saw some mountains not too far from your villages. Send your strong men and have them dig for a yellow ore. Heat up the ore and melt the metal. Pour it into molds to gain disks like this." Zuppo reached into a slot in his robe and presented a yellow coin. He put it on the table in front of Sobor. "After you've delivered to us ten buckets full of these disks, your debt will be no more."

Sobor gazed at the disk for a while and then he broke out laughing. However, he cut his laugh short, and the angry look returned to his face.

"Consider it done," he told Zuppo. "You'll have your yellow disks before winter arrives... Now, if you don't mind, I want you to leave."

"One more thing," Zuppo said. "We're taking half of your horses. You've taken many more from us. For the sake of peace, we only want half of your current herd. What's more... instead of ten buckets of yellow disks, you'll owe us only nine and a half."

Sobor wrinkled his forehead. "Nine and a half, eh?" he mumbled. "Half a bucket for the horses... That's how it works, huh! What insanity!"

"I also want our bows returned to us!" Zuppo added.

Sobor, tapping the scar on his leg, nodded. "Are you done?" he said.

"I want all of our girls and everything else we've just agreed upon on your village square as soon as possible! In the meantime, I'll talk to your people in the streets, reading what God wants them to hear... Peace to you, chief of the Korvags."

While Zuppo was completing his mission in Dudvo, Morva made himself active back home in Tolup. After one of the elders told him about the chapter of the Principium that described the route leading to the land free of Korvags, he went from house to house and told everyone to start preparing for a long journey.

Some of the Chroms had questions about the road leading to freedom.

"At the south end of Mount Ulvi, a gorge cuts through the range," Morva explained to all those who wanted to hear it. "Once we make it across that gorge, we'll continue our journey on flat land until we reach the big water at the edge of a forest. There, we build huge rafts and set sail for our new homeland."

Most of those Morva talked with liked the idea of leaving Chromland. There were some, however, who first wanted to hear what Zuppo had to say about the matter.

"Let's not rush making up our minds," suggested one elder. "Although, this plan is in the book Zuppo received from God, I suggest we wait until our leader returns."

"It's possible that Zuppo never returns," Morva argued. "The Korvags might have already found a way to destroy him. I have a hunch about this after that dream I had last night. You know what happens if Zuppo is dead. The Korvags will come and kill all of us. I say we should pack and leave."

With the exception of Morva, everyone felt relieved when Zuppo and Miran arrived at the end of the day.

Zuppo was riding back with Zorel on his side. Both of them led about a dozen of the horses. The Chrom girls, some of them half-naked, the way the Korvags chased them to the square in Dudvo, were riding the rest of the horses.

The villagers gathered at the gate to welcome the arrivals.

At sunset, Frugo was laid to rest in a brief, solemn ceremony.

The following morning, the Sevens met in full attendance.

Morva was the first to speak. He did not even wait for Zuppo to open the session as usual. "I do fear for my life," he began. "I fear for the lives of all Chroms. I'm sure I'm not alone with this concern. I'm not a coward as some of you might think. I'm a realist. The Korvags, no doubt, are animals. One day, they'll kill all of us. Now, that Zuppo has brought us the solution, we should all pack and move! Our new domain awaits us on the other side of the big water. I say let's leave immediately!"

"We'll leave when the time arrives," Zuppo responded quietly.

"We should go right away!" Morva insisted. "We can pack our essentials in a short time and load everything onto carts. We have horses! They'll make our journey so much easier. Let's not wait until the Korvags take them away again! I had that dream. I know the Korvags will come." He turned to face Zuppo. "You're the only one immune to Korvag brutality. They can't hurt you but the rest of us are vulnerable... You surprised Sobor yesterday, and that's why he gave in to all your demands. However, he might wake up tomorrow ready to attack us. I'm sure he'll figure it out… It's simple, after all. If he doesn't try to harm you, you can't disable him. You can't count on us to do it for you." He took a deep breath. "I say, let's vote on the matter."

Zuppo and the others accepted the proposal.

The outcome of the vote favored Zuppo, so they decided not to pick up camp just yet.

"It's very important that we enlighten the Korvags before we leave them behind," Zuppo said later. "We have to produce an adequate number of replicas of the Principium and deliver some to Dudvo and to the other villages so that all Korvags can learn what God expects of them. Besides, let's not forget how much they owe us! Sobor promised to repay us in Tolurs before fall is over. In the meantime, we'll send some of our men to build the rafts at the shore. They'll put up huts on those floats to protect us from the cold winds."

"Don't tell me you want us to cross the big water in the winter!" Morva interrupted.

"If you read the relevant chapter of the Principium carefully, you'll find that only the northwest winds can carry our rafts to the new land. Those winds blow only in winter."

"I see," Morva said somewhat remorsefully. "So there're things I just don't know… I should keep my mouth shut."

"Let's focus our attention to the challenges ahead of us," Zuppo continued. "We really don't have too much time before winter arrives. A lot has to be done between now and then. Let's make sure we organize ourselves, and that we don't overlook anything important."

Finally, the meeting was adjourned.

The leaders, and the assistants they selected to help them draw up and carry out the big plan, were busy all day, every day.

A small team managed to produce enough paper for a dozen books. Of course, the pages of these books were not as fine as the one Zuppo brought with him from Mount Ulvi. Nevertheless, the elders had no difficulty writing on them.

Zorel kept a few young men to hang out near the swamp day and night. Their responsibility was to report at once in case they spot a Korvag intruder. At the same time, Tolup was thoroughly combed to make sure no enemy spy was hiding in the village.

A handful of strong volunteers left on horseback to cross the gorge, find the shore, and start building the rafts. Some of them were due to return when their work was finished.

It was not before the end of the summer that a Korvag guard arrived from the other side. One of the young Chroms aimed his arrow at him as soon as he emerged from the reeds.

"Don't shoot! I'm a messenger," shouted the Korvag.

"Who sent you?"

"Our chief, Sobor… He wants to know if your people are ready to receive two of our elders whom you could teach the art of writing and reading."

"Yes," the young Chrom answered. "We're instructed to let up to four of your people to come for that purpose."

"Very well," said the Korvag. "They should be crossing over in the morning." He turned around and disappeared in the reeds.

The two Korvags, an elderly couple, were accommodated in one of the empty houses the Chroms set up earlier as a school for their own. Guards were stationed around the house to make sure the guests would not move around freely in the village. The Chroms did not want them to see any of the preparations for their planned journey.

By the time the Korvag elders learned how to write and read which did not take more than a few days, several new copies of the Principium were completed for them.

"Go in peace now," Zuppo said farewell the couple. "Take the words of God with you and first teach those who can learn fast and then teach the rest of the Korvag people."

"We appreciate the knowledge you passed on to us," the Korvag man replied. "Teaching our people won't be an easy task but we'll do our best."

"Our current leader, Sobor, has taken your demands seriously," the man's wife said. "Ever since your visit, he's been a humble man."

"The ore production in our mountains is right on schedule. We've already minted two thirds of the yellow coins we owe you."

Again, the woman took over. "There's one problem though.... Some of the men working in the mines began to grumble lately. There're rumors that they might start a rebellion to overthrow Sobor. They think our leader has turned into a pile of shit." She giggled. "Please, don't ever tell another Korvag I said this."

"Your words will stay within the walls of Tolup," Zuppo assured her. "As for the yellow coins, please tell your leader to send us what you already have. Just in case there's indeed a rebellion, let's write off two thirds of the Korvag debt."

The day after the couple's return, six buckets of the yellow metal disks were delivered to the Chroms. Miran, the bookkeeper, wrote a receipt, Zuppo signed it and gave it to the Korvag carriers to take it back to Sobor.

The news of a possible uprising in Korvagland helped to quiet those Chroms that spoke out against the exodus. Even the weak and the very old began to realize that there was no alternative.

"Our destiny is our responsibility," Zuppo repeated whenever the move came into question.

Half of the ship builders returned to the village in the middle of autumn.

"The work's done," they reported to the leaders.

Preparation for the move reached its final phase. Carts, covered with canvas, were packed with fruits, grains, beans, vegetables, and other staples. New wagons were built to carry the very young, the very old, and the ill.

"The winter is fast approaching," Zuppo told Zorel one day. "The time to leave has arrived. Send a messenger to the swamp and tell him to ask the Korvags to pay their remaining debt."

The messenger came back with troubling news.

"The Korvag guards are no longer as friendly as they used to be. They told me to go and mind my own business. Sobor is dying, they said. They also told me that there'll be war again as soon as their new leader takes over."

"Well, we won't wait for that," Zuppo said.

The leaders agreed that the remaining debt of the Korvags would have to be collected later, possibly by future generations, with the proper interest, of course.

On a nice, sunny, relatively mild morning the Chroms said good bye to their houses and all those belongings that had to be left behind. After every water-sack was filled, the caravan headed toward the mountains. There were just enough horses to pull all the wagons and carts.

They spent the first night away from home in a dried out lake. It cooled off considerably after the sun went down, so those sleeping under the starry sky, mostly the men, had to put on some extra clothing.

Bodor, one of the Sevens, had lost his mother during the summer. A falling tree, uprooted in a storm, crushed her. His father had died when he was still a child, killed by the Korvags. Now, Bodor had a new family. He and Zea, Zuppo's sister, had exchanged vows at the end of autumn.

"I hope the Korvags won't learn of our disappearance before we're safely off the land," Zea told Bodor as they moved closer to each other. "If I get captured by them once again, I'll surely die."

Bodor kissed her gently. "My guess is that currently they're quite busy with their internal affairs," he replied. "We'll see... I hope we will be boarding those rafts in time... I can imagine their astonishment when they eventually ride into our deserted village. They might think Zuppo took us all up into heaven."

Bodor and Zea were fairly comfortable on top of a cart packed with sacks of grain. They covered themselves with a soft sheepskin.

"I'm so happy to be your wife," Zea said as she wrapped her arms around her husband.

"I love you very much, Zea... I love you with all my heart. I've loved you ever since we were children. I always wanted to tell you but then, one day, the Korvags snatched you. That made me very sad. I used to cry during the nights. Now, however, I feel like I'm the luckiest man in the world."

"Oh, Bodor," Zea sighed. "I'm so glad to hear you felt the same for me as I did for you... After the Korvags enslaved me, I didn't even dare thinking of you anymore. I was so ashamed of myself."

Bodor kissed her. "Let's not talk about the past," he said softly. "Don't even think about the time you spent on the other side. It was a bad dream, nothing else. Forget it. Think of all the love and happiness we'll have together. As long as I live, I'll always love you, only you."

Zea was overcome by emotions. She had tears rolling down on her face.

"My love... you're crying?"

"Oh, no, not at all," she was laughing. "It's the happiness that brings out my tears. It's such a wonderful feeling! I've never felt this way before." She rubbed her nose against Bodor's and then, quite seriously, she added: "Thanks for restoring my faith."

One of the horses whinnied; otherwise it was a quiet night.

The following day, the convoy arrived at the gorge.

Passing through it turned out to be more difficult than they had anticipated. The ground was rocky and uneven, some of the carts and wagons tipped over. An old woman died when she fell off the wagon. The wheels crushed her to death.

The men were busy turning the vehicles back on their wheels and re-packing the loads.

Finally, they arrived at the other side of the mountain range. Another three days later, they reached the shore. The sight of the blue ocean enthralled them.

The leaders inspected the rafts that were lined up on the sandy beach.

"They look pretty impressive," Zuppo commented. "All thirty-nine of them," he added after he finished the counting.

"Nicely built, indeed," Zorel expressed his opinion. "I especially like how sturdy they look. Three layers of long, thick beams, one layer crossing the other two in between them... The beams are nicely joined, reinforced with ropes."

Zuppo turned to the man that had supervised the building. "You must've cut down half the forest around here."

"Why are the huts built up so high?" Zorel asked. "We'll have much more head room than we can use."

"If it was safe that way, we would've built them even higher," answered the supervisor. "That way, the wind would move us faster."

"What if the wind dies?"

"We may have to paddle from time to time."

"Aren't we going to lose direction?" Bodor added his inquiry.

"Don't worry," Zuppo answered. "We won't get lost as long as the sun and the stars are up in the sky. When it is overcast, the wind is our guide."

Morva also joined. "What I'd like to know is how we're going to move these monstrous structures from dry sand onto the water."

"That's not a concern," the supervisor replied. "Right now, the tide is low. In about two days, the rafts will be afloat... As you can see, all units are connected with ropes. Only the first and the last ones are anchored. Once the high tide comes in, all we have to do is cut off the anchors and off we go."

They all knew from the pages of the Principium that with steady wind it would take them no more than thirty days to reach the new land. They had plenty of food supply to last several times that long. It was the water supply that worried the leaders a bit. If they had to stay at sea twice as long as planned, they would surely use up all their drinking water. Zuppo was the only one who had no serious concerns.

"What happens to the horses?" Zea asked her husband after the inspection.

"We'll just have to let them run wild," Bodor said. "They'll survive."

The couple stood on one of the rafts, holding hands, gazing into the distance over the ocean.

"It's like a dream," Zea said with a sigh.

"Well, it's going to be quite an adventure," her man responded after a while.

"We don't even know for sure that there really is another land out there... somewhere."

Bodor put his arm on Zea's shoulder. "That's true, my darling... That's true, indeed... We are putting our lives in the hands of God. We have to have faith."

"Of course..." Zea embraced her husband. Later, somewhat mischievously, she asked: "May I tell you a secret?"

"A secret... Sure, you may. What is it?"

She reached for his hand and placed it over her belly. "You're going to be a father".

While kissing, they heard Zuppo giving the go ahead to start loading the rafts.

By the end of the following day, they moved all essentials inside the huts. Sacks of grain, covered with sheepskin, served as beds. There was little room left on the floor as each hut had to be shared by about thirty people. Some of the smaller carts were lifted onto the rafts and secured with ropes.

Shortly after sunset, a storm erupted over the land, bringing rain and wind. People sought refuge inside their new homes, the huts.

"I hope things won't get much worse after we're washed off the sand," Bodor whispered to Zea as they huddled in one of the corners.

"Can you hear it? The water dripping from the roof," Zea said quietly.

"We could even collect the rainwater when our supply runs out, provided we get rain when we need it."

Later, the rain stopped and they fell asleep.

When Zea woke up in the morning, she did not see Bodor and that got her worried. She began to feel sick to her stomach as well. She quickly got on her feet and headed for one of the doors, just in case she needed to vomit.

'Strange,' she thought as she found it difficult to balance while walking. At the door, she bumped into Bodor who was just stepping in from outside.

"Careful, darling, careful," her man said. "We no longer have solid ground beneath us."

"Really..." Zea looked outside. "Oh, water!"

"And water... and water," Bodor added. "We can't even see Chromland anymore. It's all lost in the fog." Holding Zea's hand, he stepped outside again. "Come," he said. "I'll make sure you won't slip on these wet logs."

He led her to one of the carts. "Hold on to this rod," he advised her. "I'll do the same. The wind is not terribly strong so we should be safe out here... See? Some of the rafts are touching. The wind must have pushed them together."

"I'm almost scared," Zea admitted.

"We'll be just fine. Do not worry!"

"As long as you're by my side..."

A strong gust tore a loose piece of plywood off the side of the cart and hurled it up into the air. It flew over Zea's head, barely missing her.

"Well, we'd better get inside!" Bodor said in a big hurry. "The storm seems to be getting stronger."

"Look at that wave!" Zea cried out. "Oh, and can you see that other raft? Look! It's breaking away fast."

"I see," Bodor replied.

"Look at that!" Zea shouted. "The rope just snapped."

"Yes, it did… Hurry! Hold on to my hand! Let's get in there before we get washed off by the waves!"

Once inside, back safely in their corner, they began to wonder what they really got themselves into.

Part 2

(About 3,000 years later)

Shrigo, the fourteen-year old son of Morten, the trader, stopped at the pier on his way home from school. He enjoyed taking this route, especially when the evenings were mild.

It was the middle of summer.

Shrigo decided to miss dinner for a change. He was getting somewhat fed up with the rigorous schedule his parents tailored for him. It was school in the morning with the rest of the city kids and then school again in the afternoon at Gumiel's house.

He sat at the edge of the pier, hanging his legs, and watched the setting sun dipping into the ocean.

It was already dark when he left the pier and headed home.

When he turned into the long, narrow street, named after the ancient saint, Zorel, he heard a short, sharp whistle coming from behind the old church building. He slowed down under a streetlight and looked around. He walked on trying to keep his steps quiet on the cobblestones. He lived at the other end of the street, more than a thousand feet away.

Shrigo felt pretty uneasy, especially because the street was deserted. He had already passed the church when he heard another whistle. Shame or not, he started running.

Suddenly, two boys appeared from an arcade. One of them grabbed Shrigo by his arm and dragged him to the ground. The other one started kicking his head.

The two boys were older and stronger than Shrigo, and they were in masks.

Since they acted without saying a word, Shrigo had no idea of their identities. He defended himself as much as he could. He covered his face with his school bag.

After several nasty kicks and punches, the attackers forced Shrigo to lie flat on his stomach. One of them pulled a white chalk from his pocket and marked a big letter C on Shrigo's red shirt. The two boys then ran away and disappeared behind the church.

Tapping the bumps on his head, Shrigo stood up and dusted his pants. His fear was gone. In fact, at least for a brief moment, he was inclined to run after his attackers.

"Oh, no.... Not this!" Numia cried out when she saw her son's bruised face. "You got beaten, didn't you?"

Shrigo nodded, licking his bleeding upper lip.

"Who did it?" his father asked.

The boy shrugged. "How would I know? They wore masks... Two boys... It happened in a darker stretch of the street anyway."

Shrigo's sister, Trimea, entered the living room. She, at age nineteen, was five years older than Shrigo. After just taking a bath, she was in a robe, her hair wrapped into a towel.

"What's that on your back?" she asked startled. "Mam, Pap! Come and see! It's a big letter C."

"Of course," Morten said, shaking his head.

"Of course?" responded the girl. "What do you mean?"

Numia looked at her husband.

"Well, I guess, you two should know," the man began.

"Know what?" the siblings urged him.

"Know that beatings are now happening more and more often... It's probably the first time here in Uronton. However, we've heard that in other cities, especially abroad, the beatings have become more widespread. In Mitland, where the Chrom population is no more than one tenth of the Mit people, standing at about half a million, a beating takes place almost every day. Of course, it's always the Mits harassing the Chroms and not the other way around."

"But why... I mean why the beatings?" Shrigo reacted angrily.

No one answered.

"Just be careful from now on," his father said eventually. "Avoid being alone, especially at night. Best if you always get home before dark."

Someone was knocking.

"Oh, it must be Grog," Trimea said, closing the robe over her breasts. "I'm coming!"

"Better not to say anything about this incident!" Morten instructed the family. "Shrigo, go and wash your face! We'll tell Grog you fell down the stairs... Better yet, stay in your room while he's here!"

A handsome, young man entered. He was somewhat taller than Trimea. He had black hair, fair complexion, and brown eyes. He was nicely dressed, wearing beige leather shoes, a pair of white slacks, and a deep purple T-shirt.

"Thanks for wearing my favorite colors," Trimea said as she kissed Grog's chin. "I'll get ready in a hurry. Have a seat."

"You don't have to hurry," Grog replied. "We have plenty of time until the play begins."

"Good evening Morten, good evening Numia," he greeted the parents after Trimea slipped into her room.

"Nice to see you, Grog," Morten said. "How's school?"

"I'm doing all right, I guess... I've just passed a couple of important exam. I passed them with excellent marks."

"Good... That's really good." Morten was struggling to find something else to say. He looked at his wife.

"Oh, I guess... you will then soon finish your studies... and get a diploma," the woman took over.

"It won't take too long now. I'll be happy when the cruel years of school are finally over. I'm getting tired of books and that's a fact."

"The work gets even harder once you're out of school," Morten said. "Obviously, you'll have a job."

"If I find one," the young man shrugged. "Well, I can always help my father on the farm, I guess."

The woman answered: "You'll find a position, I'm quite sure of that. Accountants are in short supply.... Isn't that right, Morten?"

The husband agreed. "In fact," he added, "my friend, Tercal, will be looking for someone soon. His bank keeps growing. He's going to open a new branch on the other side of the river."

After Trimea got dressed and the youngsters left, Numia put her arms around Morten and rested her head on the man's shoulder. "We've lived in relative peace for a long time," she said after a sigh. "I'm wondering what the future holds for us... People all over are getting restless... What kind of world our children and grandchildren are going to live in?"

Morten kissed Numia's forehead.

"Let's face it, darling," he said quietly. "We aren't the only smart folks anymore. Many of the non-Chroms are catching up. They're not dumb, and they also want a piece of the pie."

The woman glanced at him from the corner of her eyes.

"Well, we can't let them have it, can we?"

Morten nodded. "You're absolutely right. We must draw the line somewhere… Otherwise, they'll take over."

"We don't want that to happen, do we?"

"No, we don't, my dear. In fact, we won't let that happen. No need to worry. Plans are being drawn up to remedy the situation as we speak."

"Is that why those two eminent Chroms from Mitland are visiting here in the near future? Gosh! I keep forgetting their names. I know one of them is a writer."

"His name's Krupchek. His comrade, Shrolen, is a banker. Not long ago, they met some very influential businessmen in Esiland. Chroms from as far as Murumbia were also attending. The entire conference was disguised as a birthday party. I hope the Eslans didn't get suspicious."

"Oh, the Eslans," Numia shook her head. "They're worse than the rest, aren't they? Isn't there some Korvag blood in those bastards?"

"There might be… Anyway, I'll be quite busy when Krupchek and Shrolen arrive. This upcoming conference is a very important one. We'll be designing some fundamental changes."

The woman yawned.

"Perhaps we should just go to bed," proposed the man.

"Shrigo," Numia shouted.

"Yes, Mam," the boy answered from his room.

"Your dinner's on the table. Warm it up! It's probably cold by now."

"I'm thinking about this Uron fellow, Grog," Morten said while they were getting undressed. "Quite a dilemma…"

"You don't think it's a good idea, do you?"

"Well, if Trimea loves him, there's not much we can do. It's her decision… I trust her. I'm sure she knows what she's doing."

"There have been a few mixed marriages that worked out just fine… Grog isn't a dominating type. Trimea should be able to handle him."

"Well, fresh blood… We'll just have to keep an eye on the grandchildren."

Every fifth day, on Zumdays, the bells of the churches on the continent reminded members of the Zupolich faith that it was again time to attend the service. There were two masses, one in the morning and one late in the afternoon.

There were exactly twenty Zumdays in each of the four seasons. On Zumdays, nobody worked.

In Uronton, it was Mengor who greeted the faithful as they entered the church.

Mengor, the Zupolich priest was an old man. His father and his grandfather, even his great grandfather, they were all Zupolich priests.

Mengor never got married. According to some, he lost his reproductive organ when he was a child, falling from a tree, getting injured by the remains of a broken branch. Anyone looking at his shriveled face could tell he was still feeling the pain of that horrible experience.

It used to be that all inhabitants of the planet, not counting the Chroms and a few wild natives of the third continent, were of the Zupolich faith. It had been that way for centuries after followers of the saints finally convinced people that the Principium was God's teaching.

The Chroms never joined.

A few years back, a brave young man by the name Rhodi started a revolution. He and his followers broke away from the church and created a new denomination they called Reformed Zupolicism. In the beginning, the High Priest in Zupcan, the city-state that controlled Zupolich affairs, had Rhodi prosecuted. Later, however, since the number of renegades remained small, the prosecution stopped.

Rhodi was a native of Uronton. His tiny congregation usually met on the eve of Zumday.

Once Rhodi's followers built a small church on the outskirts of the city but when it was completed the Zupolichs demolished it. So, they continued meeting at Rhodi's house.

Rhodi did not reject the Principium in its entirety but he wanted more freedom to think. "I didn't mind sending some of my money to Zupcan," he often told his congregation. "What I disliked was that the High Priest and his circle also demanded my soul as well as my mind."

The Chroms being a secretive, separate bunch also had a lot to do with Rhodi's rebellion. He and some of his closest friends often had discussions about the Exploiters as they referred to the Chrom people.

"Answer me this," Rhodi said on one occasion. "How come these Chroms are allowed to go their own way? I haven't heard that Zupcan has ever tried to convert them. Just think about it! Zumaniel, the Chrom, received the Principium from God some three thousand years ago. Zumaniel's disciples, the saints as we call them today, began converting the world. The rest of the Chroms, however, rejected Zumaniel, saying that the Prophet forced them to cross the ocean on fragile rafts, resulting in half their population dying by the time they reached this continent. Today's Chroms are not a bit more lenient when it comes to condemning Zumaniel. They say it's the Prophet's fault that they're scattered all over... The Addendum tells us how the storm separated the rafts of the ancient ones, sending some of them as far as Feroland in the south hemisphere. Well, perhaps those that didn't survive the crossing could've blamed Zumaniel. The ones that made it should've been grateful. He delivered them from the claws of the Korvags, didn't he? Today's Chroms get me really confused, I must admit. They found peace and prosperity on our continent. They're much better off than the large majority of the non-Chroms. They control the money and that alone makes them very powerful. What are they so unhappy about? They're the ones who should really worship Zumaniel! Don't you think?"

His friends all agreed.

"As I see it, there're two problems in our world today," he continued. "One is Zupcan and the other one is the Chroms. We rebelled against the High Priest but how do we challenge the Chroms?"

"They do get beaten from time to time," someone said, making the whole group laugh heartily.

"That'll just make them stronger," Rhodi replied. "They'll probably appeal for sympathy and the Zupolich sheep will feel sorry for them."

"Let's do something," one of his friends suggested.

Rhodi shook his head. "We're non-violent people," he answered. "Unless we resort to aggression, what'll probably never happen, there's nothing we can do."

They left it at that.

When Rhodi and his followers were in better mood, they enjoyed joking about Mengor.

The old priest was a compassionate man. He knew the personal problems of every Zupolich in his church. He also knew how to console those who suffered mentally or emotionally. Generally, he was well liked. He had no difficulty keeping his herd together and that was a big thorn in Rhodi's eyes.

On the tenth Zumday of summer, Mengor was in for a surprise. He saw a Chrom enter his church. It was Trimea, the Chrom merchant's daughter, arriving hand in hand with Grog. What's more, the Chrom girl stayed for the entire mass.

The whole city talked about it for days.

"Perhaps they're ready to repent after all," one woman told her husband.

"I don't believe that for a moment," the man replied. "She's in love, that's all. She'll do anything to corrupt that Grog boy. Once he's hooked, you won't see her in our church any more. You have my word on that."

On the following Zumday, as Grog was getting ready for church, his father wanted to have a talk with him.

"The folks are telling me that you go to church with a Chrom girl. Is that true?"

"Yes, father, why are you asking?"

"Why am I asking?" The man raised his eyebrows. "Well, I'm asking because you're my son.... I've been feeding you, clothing you, and even sending you to school. Don't I have the right to ask?"

"Of course, you do, father... I just..."

"Look! I don't mind if you go to church with a girl. You're not a small boy anymore. You grew up and it's time for you to find yourself a wife." Suddenly, he started shouting: "Why a Chrom? Why? That's what I'd like to know! There're plenty of nice and decent girls around. Aren't there? Why do you have to bump into a Chrom? Why? Tell me!"

"Father...."

"Listen to me, son! The Chroms are a weird people. Everyone knows that... They're a devilish kind. A handful of them outsmart an entire city. I always have goose bumps on my arms when one of them looks at me.... Listen to me, Grog! You don't belong with them!"

"But.... I love her," the youngster pleaded.

"You just happened to run into the wrong female, that's it. Go and find yourself another one. Find yourself one of your own! You would never be happy with a Chrom. Let me tell you a story. This happened before you were even born. I had a good friend when I was about your age. He, just like you, fell in love with a very pretty Chrom girl. He married her in the spring. On a summer day, he was dead. He killed himself. Do you want to know why? Well, I tell you why... He found her in bed with her own brother. That's why."

After a pause, he continued. "We hear all kinds of crazy things about them. If only a tenth of that's true, it's already too bad."

"Father, I know Trimea very well. We went to school together."

"You'll never know a Chrom. You might think you do but you don't. You just don't... I'm telling you, their screwy mind is beyond our comprehension."

"I understand Trimea..."

"Why are you so stubborn, son? Why? Listen to your father! I don't want to hurt you... I only want to save you from hell."

"Hell? Why do you..."

"Why? Well, I guess I haven't said enough yet. Well then, it's time for you to know what really happened to your mother... One of those Chroms running around with those little leather bags in their hands... those Chroms that began popping up everywhere when I was still a child. They called themselves medics. Well, one of those bastards killed her. Of course, no one could say that is was really murder. Even I had to keep my mouth shut. It happened when you were born. Your mother had a hard time delivering you. That medic told me that the only way was to cut her belly open. He called it an operation. He said her wound would heal and she'd live. Well, she didn't. He saved you but sent her to the grave."

Grog looked at his father who was struggling to hold back his tears, and then he said he was late and left. Trimea was already waiting for him at the church.

A few days later, Grog was murdered in the street. An old woman found him in the morning with a long knife in his chest. Some speculated that it was his father who hired the killer.

Flukk, head of all Nobles, the highest ruler of Murumbia, took his seat at the head of the table. The two bodyguards that escorted him to the dining chamber discretely retreated and waited outside the door.

Flukk's guests, about a dozen of the Nobles from the most important cities of Murumbia, had been sitting around the long table for some time, waiting for their superior to arrive.

"Thank you for coming," Flukk greeted his guests. "I hope you enjoyed the journey across the various parts of our great domain. I'll see to it that you're comfortable in my castle as long as you stay. Let me know if you don't like the quarters provided for your accommodation, I'll relocate you."

After the greeting and some polite conversation, dinner was served. When they finished eating, Flukk invited his guests to join him in the conference chamber.

"Our worst fears seem to be coming true," the Head Noble began after the group settled in around the huge oak table. "The situation has definitely got more dangerous since we last met here a year ago. At least that's what I see here in the capital city. Rivertown is full of rumors. My people are complaining more than ever before. Of course, I need to hear what you have to say. I would hate to jump to conclusions based on my own beliefs. After all, it's possible that the people exaggerate. So why don't we go around and hear what all of you have to say." He looked at an old, gray-haired man on his right. "Go ahead Murok. You're the oldest and possibly the wisest Noble in all of Murumbia. How do you see it?"

"Respected Ruler," said the old man, briefly bowing his head toward Flukk before carrying his glance around the table. "My dear fellow Nobles... I'm so glad we're meeting here today. I've been anxiously waiting for the opportunity to report to you what I think is the most alarming news. In my city, and in the whole of the province under my jurisdiction, the activities of the Chroms have created a situation that, in my humble opinion, calls for immediate action on our part. Let me put it the way it is. The treasury is no longer the mightiest power. The Chroms have accumulated wealth that now dwarfs the treasury. This alone is a great enough threat. Unfortunately, this is not all."

Murok took a sip of his wine licking his lips, and then he continued:

"Since our last conference, the number of Chroms becoming independent has at least quadrupled. The trend is accelerating. Now, only a handful of them live off the land. They keep opening up businesses. Where the hell they get their satanic ideas from, I have no clue. Just recently, they bought up a number of small houses in the heart of our city.

One of them, son of that banker, Tivos, established what he calls a real estate company. Using the profit from his father's bank, I suppose, he hired a bunch of peasants to tear down those houses and build a... spital or whatever the hell he calls it. Can you believe it? First the high school, now a spital... In that school of theirs they produce medic after medic. Now they need a spital where all those medics can experiment. They say they know how to make people live longer. Guess what! Our people naively believe them. More and more of the Sivors stop using those good old herbal remedies that have kept us in relatively good health for as long as we can remember."

Almost out of breath, the old man paused.

"Of course," he went on, "I don't have to tell you that hardly any of the Sivor youngsters can enroll in the high school to become medics. First of all, the Chroms charge a high tuition an ordinary Sivor can't afford. Then they have what they refer to as entry examination to filter out those they don't want to see enrolled. Strictly out of curiosity, I sent my grandson to take the test. Sure enough, he didn't pass. He scored perfect on the written test but the oral examination eliminated him. I'm telling you, the Chroms are taking over our society. We must do something before too late... It's not enough that one of them invented that darn steam engine to power ships and trains, and to produce artificial light, now they're talking about engineering machines that'll fly. What's this world coming to? Of course, they keep increasing their influence as they grow their financial strength. Now they own the railroads, they own the steam liners and they own the power stations... Do I have to continue? Soon they'll tell us what to do."

Another one of the regional rulers spoke even more bitterly: "All of the Chrom businessmen now have at least two bodyguards. They can pay more than I can. I have lost a few of my best men in recent days. That's not all though. There're rumors that soon there'll be a Chrom police force and a Chrom army made up of mercenaries. Before we know it, they'll be sending their army to defeat ours. That's going to be the end of us. I don't hesitate to say that I believe they'll kill all of us."

After all the reports, the mood in the chamber turned extremely anti-Chrom.

"If you don't mind me summing up the situation," Flukk took over again, "Murumbia is in grave danger. We must act, and we must act now! The question remaining is this: what can we do? I have some ideas of my own but I'd rather listen to your suggestions first. Feel free to recommend any radical measures. We'll weigh them and then we'll vote on them."

One of the Nobles went as far as suggesting that all Chroms over the age of fifteen be put in prison for the rest of their lives.

The group finally agreed on an equally drastic but less inhuman solution.

"We'll give them exactly one year," Flukk said after the vote. "That should be a reasonable length of time for the last Chrom to leave Murumbia. Any of them caught within the boundaries of our country after that deadline will be executed. Where will they go? It doesn't matter to us. They can go to Feroland, the country to the west of ours. The Gombis people might even welcome them. After all, Feroland is about the size of Murumbia with only half the population. Hardly any Chroms..."

"Chroms don't like to live in the desert," someone commented.

"That's not our problem," Flukk continued. "They should be glad we let them go in peace. If they don't like Feroland, they should go to Sutronia, our northern neighbor."

"That might just be too hot for them. It's too close to the equator," another one commented.

"All right, then let them go to the third continent! Khamiria is almost uninhabited. A couple of primitive Khamir tribes, that's about it. The Chroms can enslave them if they want to. As for the move, they crossed the ocean once before. All they had were rafts. Today, crossing isn't a problem. If they don't have enough ships of their own, we'll make ours available to them. We won't even charge them."

At the end of the session, Flukk called in his secretary to write the Decretum.

"This new law is in effect as of tomorrow," Flukk announced after all Nobles signed the document. "My librarian will make replicas of this document so that all of you can take it with you back to your provinces. Send your messengers even to the smallest villages to make sure everyone learns of our decision. Our armies will stay on high alert until the last Chrom is gone."

"Oh, Ravar, it felt so good," whispered Lella kissing the man's earlobe. "Now we're husband and wife… and I assure you that I'm the happiest of all wives."

"You're wonderful," the man responded in a gentle voice. He caressed Lella with both hands.

"I love you, my dear man." Lella pressed her mouth against Ravar's for a moment. "I really, really love you." Then, playfully, she jumped off the bed and stretched her arms. "It's time to have a sip of that Amoritto liquor," she said winking at Ravar.

"I agree," replied the man, licking his lips. "I'll get it right away. The processor will prepare it in no time."

"Careful!" Lella warned him when he got on his feet. "Remember, we're in my sleeping chamber, and it's hardly your size."

Ravar laughed. "I know. I programmed it for you."

"Wow!"

"Easy… Look!" He pressed a few numeric keys on the wall.

"Wow! The ceiling has just moved up a few inches," Lella said. Smiling seductively, she added: "Are you going to leave it that way?"

"Well, if we're to spend more time together...

"That's what I mean". She kissed him again. "Go my love and get that liquor, will you?"

After they both emptied their glasses, Lella sighed.

"Now that I'm potentially pregnant, I wish more than ever that we had solid ground under our feet," she said.

Ravar nodded. "I'm with you on that one hundred percent."

"So, how's everything down there?"

Ravar paused and invited Lella to sit next to him on the bed. "We're right on schedule," he started. "In fact, we're probably ahead of schedule. Wrix has been doing a fantastic job. The bioblaster has had its effect as well. The mental capacity of the Chroms keeps improving. We've only been here for a short time, and already they're designing flying machines. Every generation gets sharper than the previous one. They've really created a lot of great things already. Of course, then there're those who try to slow down the progress. I'm referring to the non-Chroms, as you might have guessed. Many of them are stubbornly against going forward." He paused again. "The Chroms... They're really great! We couldn't have hoped for anything more. There're so few of them compared to the rest of the population, still they've managed to gain control just about everywhere. However..."

"What?" Lella's eyes opened wide.

"Well, I guess I ought to tell you..."

"Is something wrong?"

"While you were sleeping last night, the computer reported a significant crack in the outer shield. I was able to seal it with the help of the robots before it was too late. We're safe now but it's possible that additional cracks will develop."

"What then?"

"We could lose our oxygen if the repair is not done right away."

"Can't you program the robots to do whatever is necessary?"

"I'd have to monitor the repair to make sure it's done perfectly right."

"Well, I sure hope to be on the surface before another problem develops."

"Lella, believe me, I'm doing my best to get us down there. In fact, I'm going to speed up the process."

"Speed it up? I wonder how?"

"Increasing the intensity of the bioblaster to the max..."

"If I remember well, you told me that the Chroms could get suspicious if suddenly they became too clever?"

"Guess what! At this point, it doesn't matter anymore. They like the position they are in, I'm sure they'd hate to lose control. They're no longer nomadic people, and so decency is the last thing on their mind. The planet's population has been increasing like crazy and they understand what that means. Eat or be eaten... Don't worry. I know what I'm doing."

"What about those other creatures? You know, the non-Chroms... How are they coping with all these changes?"

"Naturally, friction is unavoidable... Of course, if wars break out, the Chroms will benefit. They'll just have to find a way to direct the aggressive behavior of the non-Chroms against their own kind... As for Wrix... it is time for him to assume a much more important role. I will also allow him to improvise and act more freely."

Lella smiled. "I see you've got it all worked out, haven't you?"

"You bet, my darling."

Lella kissed him. "I've always known you were one of the sharpest minds in our Galaxy. I love you! I love you! I love you!" With that said, she aggressively climbed over him, wrapping her arms and legs around his naked body.

Rhodi and some of his followers sat in the shade of a large tree on a hillside. It was Zumday afternoon.

"We should come here more often," a young man with a long beard commented. "Especially on nice summer days like this one... Out here it's not hot, and the sky is so beautifully blue."

"We can really enjoy the gorgeous view of the city from up here," someone else said.

Later, when Rhodi spoke, everyone paid close attention to him.

"I may shock you with what I have to say today," he began. "First of all, and I'm not asking for your pity or for any condolences, I have to inform you that a few days ago someone attacked me in the park. As usually, I stayed longer than anyone else. It was already semi-dark between those chestnut trees when I decided to rise from my favorite bench. I only made a few steps when I was hit in the head from behind. I fell down and passed out for a short while. My attacker, or attackers, must've thought I was finished. I was lucky they took off right away... My skull is not broken, and the bruise is going to heal soon."

He took a long, deep breath before he continued.

"The point I'm trying to make is that it seems we can no longer live the peaceful, undisturbed life we've been taking for granted for quite some time now. It's sad. It's very sad... Now, I don't know what the intention of my attacker was, I can only guess. Everyone knows I'm a poor man, so robbery is out of the question. I don't remember ever insulting or intentionally offending anyone in this city. That's, of course, not to say there're no people who hate me. After all, those with hatred in their hearts will always manage to find targets for themselves."

He paused again.

"Still, I refuse to believe that negative emotions alone would prompt someone to kill. Of course, it's also possible that the attacker didn't mean to kill me. Perhaps the blow was meant as a warming. This seems like the most logical explanation." He carried his glance around without looking at anyone in particular. "It's not entirely impossible, either, that someone from our congregation spies on us. That person may even be here right now. Frankly, I couldn't care less. In fact, I'd be thrilled to know that whatever we discuss during our meetings gets to the ears of those afraid of our movement. After all, we have no bad intentions. We don't want to harm anyone. Now, if they're scared because we speak the truth, that's their problem. All we want is basic rights for every single individual. Anyone disagree with me on this?"

"No, you're absolutely right," said the bearded youngster.

Everyone else expressed similar approvals.

"It maybe that we'll have to start living in fear after all," continued Rhodi. "Especially, if the following rumor proves to be true... I know a few people in Esiland who speak our language very well. They're pure Eslans, or so they say. I received a letter from one of them a couple of days ago. In that letter, it's written that many Eslans came to a frightening conclusion. According to that conclusion, the High Priest in Zupcan is a descendant of the Chroms."

"What?" The young man sounded startled. "The High Priest of the Zupolichs is a Chrom? Wow!"

"Well, it doesn't sound like a very illogical conclusion to me," an older man commented. "However, can this be proven?"

Rhodi shook his head. "I don't see how anyone would go about trying to prove it. Nevertheless, it's interesting to consider the possibility. Now, if the High Priest really is a Chrom, then probably all or most of the previous High Priests had also been Chroms. It's not unlikely that Zupcan is the center of a tree thousand-year old conspiracy. Let's look at reality for a moment! The Chroms, after arriving on our continent, faced the enormous challenge of integration. They were a very clever bunch. They could've easily worked out the following strategy. I emphasize, this isn't proven, or at least not yet... Now, if any of you don't want me to go on just say so."

"Let's hear it! Let's hear it!"

"Okay... Feel free to interrupt me anytime with comments, questions or objections." Rhodi briefly looked at the grass in front of his shoes before he continued. "It's possible that after the Chroms arrived, some of Zumaniel's disciples formed a separate group, went underground, and managed to assume non-Chrom identities. Let's call the members of this group the Implants. Of course, the Implants simply carried out the orders of the Chrom leaders. At the same time, in a cleverly orchestrated theatrical act, the Chrom population abandoned Zumaniel and the saints. As the Prophet and his disciples, the saints, began to spread the word of God, the Implants acted as converts. Later, more and more of our ancestors followed the Implants joining what is today the Zupolich church."

Rhodi stopped. For a few moments, he searched the faces of his listeners.

"The Chroms knew how to employ reversed psychology that was probably even more effective three thousand years ago than it is today. The more they condemned Zumaniel and his teachings, the easier it was for the Implants to convert non-Chroms to the Zupolich faith... Now, another interesting fact is that for quite some time after the Chroms' arrival, Zupolicism was limited to our ancient Uronia. Why? It's because only the ancient Uronian language was very similar to the Chrom language. It was not difficult for the Implants to infiltrate Uronia. Whenever they moved into a village, their language was mistaken for a Uronian dialect from another part of the land. Later, some Implants moved to Mitland, disguising themselves as Urons. None of the Mits ever suspected that they were Chroms. Anyhow, in just a couple of generations, the Implants were all over. Anywhere they moved, when asked, they said they were of Uron descent. They continually played an active role in society. They were generally well liked as they diligently praised Zumaniel and condemned the Chroms."

While Rhodi spoke, he often rested his gaze on the young man's beard.

"During those times, the Chroms, I mean those who didn't conceal their real identities, also moved around, eventually showing up just about everywhere. They maintained their secretive way of life. They developed a strange dialect of their language that not even the Urons could understand. They spoke in that tongue every time they didn't want the locals to understand them. Any wonder people hated them? As I understand, the Implants are also perfect in that dialect but they never use it when dealing with non-Chroms. Of course, the dialect can also be learned. We know that some non-Chrom scholars have been fluent in it. Well, I now suspect that most of those non-Chrom scholars have been Implants... As the Implants appeared to be faithful Zupolichs, they had no difficulty in becoming the leaders, the teachers, or the priests of society. Their main role, of course, was in the church."

"The priests made sure that the masses followed the directives of the Principium. They made the people believe that if they don't do what the sinners do, and if they forgive the sinners their sins, even if those sins were to take advantage of them, they go to heaven after they die. Who were the sinners? The Chroms, of course... The priests never hesitated to point that out. Clever, isn't it?"

"Wow!" The young man nervously played with his beard. "Are you actually suggesting that the Principium is not really God's words? That it was not only written by but is also the brainchild of the Chroms?"

Rhodi frowned. "Well," he said, "I'd rather let you draw your own conclusion."

"I see," the young man answered quietly, rubbing his forehead with one of his fingertips.

"Go on Rhodi! Go on!" somebody urged him.

"There's not much more to it, really," replied Rhodi. "We more or less know the rest. We see that the priests have been doing a real effective job on the masses. This allowed the Chroms to exploit the Zupolichs and build their financial empire. The Zupolichs have been aware of the exploitation but continually forgave the Chroms because that's what God expected of them, according to what the priests made them believe." He nodded. "I'm telling you, it's nothing short of a genius plan. Then, there's Zupcan, of course. The Implants, alias Zupolich priests, established the city-state to protect the faithful. What was the price of the protection? Oh, only a tenth every payday... Look at Zupcan! It's wealthy beyond belief. In exchange for all those tithes, it keeps feeding the masses to the Chrom sharks... I'm telling you, it's not a very promising situation. It appears, or at least we can now suspect, that the governors in most of the cities all over the world are all Implants. The lands that most likely are still governed by true locals are Esiland and Murumbia."

"So, where does all this leave us?" an older man asked.

Rhodi shook his head. "I don't know... I really don't know." Pointing to the western horizon, he added: "What I do know is though that those fast approaching dark clouds will get us soaking wet if we don't pick up and leave right away. I suggest we eat our sandwiches when we get home."

All of them left in a hurry.

The next day, Rhodi disappeared. No one ever saw him or heard from him again.

Wrix leaned against a rail on the upper deck of the steamship named after the famous Sivor explorer, Margas. He was dressed warm, like all the other passengers, as it was still winter in the Southern Hemisphere.

The Margas left Rivertown harbor in the morning. It was already late afternoon when it cleared the Cape of Murumbia, the easternmost point of the entire continent.

"What a magnificent sunset!" Wrix heard passers-bye saying while the ship distanced away from the land.

Wrix agreed. He watched the deep orange star setting on the horizon.

"Magnificent, indeed," he whispered to himself.

A pretty, young woman joined him. "Expecting someone?" she asked as she casually leaned on the rail.

"Sorry?" Wrix responded as if he was surprised. "Ah, you mean..."

"Exactly!" the woman interrupted him cheerfully. "I'm wondering whether you're traveling alone."

"Oh!" Wrix cleared his throat while taking a better look at the woman. 'If I was real, the way I appear, I could very well be her father,' he thought.

The woman laughed. "All right, darling, you don't have to tell me. Your wife is probably down in the bar, flirting with one of the crewmen. Never mind! I'll help you forget your sorrow."

"Please!" Wrix said in a hurry when the woman moved closer to him.

"Not here? Well, I can understand that. Your wife might see us and kill you, right?"

"Not exactly..."

"So then, what's the problem? You know? You actually impress me. And don't think for a second that I'm one of those street girls. I just behave like that... today, because I had a bit too much to drink. But really, those girls can't afford paying for a trip on an ocean liner like this. In fact, they probably can't afford a trip on any ships. My father is rich... and I really like you. It must be those shiny lips of yours... or I don't know. I'm just getting crazier and crazier about you." She tried to put her arm on Wrix's shoulder.

Wrix stepped back. "Look!" he said in a cold voice. "I don't mean to spoil your evening but... but I'm sick. I'm very sick."

"Sick?" The woman raised her voice as she giggled. "You look as healthy as one can get."

"I really mean it!" Wrix insisted. "Actually, I'm probably terminally ill. This trip is my last hope. I heard of a medic in Landis who can operate on patients with the problem I have."

"What problem is that?" The woman was finally taking Wrix seriously.

"I have a large, ulcerous growth in my stomach. Every morning I throw up a lot of bloody pus."

"Oh, gross!" The woman grimaced. "Stop, don't even tell me any more! Please! I'll leave you right alone... Hope your operation will go all right."

Wrix smiled as he watched the woman going away, putting her hand over her mouth. After casting another glance at the glowing horizon, he walked toward the stairs leading to the cabins on the lower decks.

"Seven... eight... nine," he read the numbers on the doors. Finally, he stopped. "Well, this is it."

"Who is there?" he heard the man's voice from inside after he knocked on the door.

"It's an old friend," Wrix replied.

The man opened the door and peeked out.

"An old friend?" he said nervously. "I don't know you. Who are you?"

"My name is Wrix. We need to talk. It's really very important."

The man hesitated. "Do you know who I am?" he asked.

"Yes, of course," Wrix replied smiling. "Your name is Diavo. You live in the center of Rivertown. You own two banks and a lot of land... and you're a Chrom. Your grandparents moved to Murumbia from Uronia exactly..."

The man raised his hand. "Okay! Okay! I'll get dressed and meet you in the coffee bar. That's a quiet enough place. We can talk there."

Wrix shook his head. "I have a better idea," he said. "You won't need to get dressed... I mean right now you are just fine the way you are in your pajamas. We'll talk right here, inside your cabin." He gently pushed the door open and stepped in. Once inside, he locked the door.

"What do you want from me?" the banker asked, looking at Wrix rather uneasily.

"Relax!" Wrix said as he looked around. "Nice, comfortable cabin you have."

"Well, would you like to sit? These armchairs are pretty soft. I'll get you something to drink. What would you like?"

"Anything… Water, get me some water, please."

When the man returned with the glasses, he seemed much less nervous. In fact, he appeared somewhat curious.

"Am I supposed to know you?" he asked as he handed the water to Wrix.

"You will," Wrix replied after taking a sip.

"You will… What's that supposed to mean?"

Wrix opened his arms. "You will… Whatever that means… At the end."

Diavo shook his head. "I'm not sure I'm following you… What else do you know about me?"

"You're going to an important meeting disguised as an educational conference," Wrix said bluntly.

The banker's face turned pale. "That's true," he stuttered. "How would you know? Who are you?"

"I've just told you, my name is Wrix. Take it easy, Diavo. You're in good company. If you really want me to, I'll tell you more about myself. However, right now we have to talk about you. After all, you're the one with the important mission."

"Okay! What do you need to know?"

"Diavo, I know you've been a very successful man. Tell me something! Do you believe in God?"

The banker's eyes widened.

"Do I believe in God?" he repeated the question. "I guess I do."

"Are you afraid of what awaits you after your life on this planet is over?"

"You know, Wrix… Am I saying your name right? Okay! So, what I am trying to say is that you're making me very curious. Perhaps you're here to tell me a few things I don't know?"

"Perhaps… So, Diavo, are you afraid of dying? Whatever that might mean to you…"

"It's an interesting subject, to say the least."

The banker reached for his robe that was lying on his bed. "I am almost shivering. I must put this on… As for your question, I'm not sure I have a clear answer. I don't think anyone really knows what happens after we die. I sure would like to believe that it doesn't matter when it happens."

"You're on the right track, Diavo. You'll make this much easier for me than I thought."

"Now, wait a minute!" the banker protested.

"Relax, my friend! There's no need to panic. Just sit back and listen to what I have to say."

"Oh, okay," Diavo responded leaning back.

"So, let's go to the heart of the matter," Wrix said with a serious expression on his face. "You, Diavo, one of the richest and most influential Chroms in Murumbia, are traveling to Uronia. Your plan is to get off this ship at Costis, the harbor town just south of the Cape of Esiland, then take a train through Landis, Mitopolis and Zupcan to Uronton. In Uronton, you are to meet with a dozen or so of the top Chrom minds to discuss the current situation as well as to work out a plan to assure the survival of your people."

Diavo shook his head in disbelief. "How do you know all this? Don't tell me there's a spy among the Chroms because I won't believe it."

"I assure you, the information is of divine origin."

"What?"

"Listen, Diavo! It's time for me to reveal who I really am... You know the story of Zumaniel, right? You do? Okay... Then you also know that God appeared to Zumaniel in the cave."

"Wait! You! You're that... creature, right?"

Wrix frowned. "Creature... Well, an interesting description. It's fine with me if that's the way you like it."

Diavo shook his head again. "Well, I hear what you are saying but... what if you're just a well-informed mortal?"

"Watch, Diavo!"

"Oh! My..." That was all the banker could say when he saw Wrix's head taking on a brand new appearance in less time than he needed to blink.

"So, how do you like looking at your own face?" Wrix said. "I did my best to come up with an exact replica."

Diavo was truly stupefied. He was unable to squeeze out a single word for some time. Finally, he covered his eyes and said: "Could you... could you just go back to what you were before?"

"No problem," replied Wrix. "You can open your eyes now."

"Unbelievable," whispered the banker.

"Okay, let me go on..."

"I know!" Diavo interrupted. "You want me to be the new Prophet."

"Not exactly... Please, hear me out. Here's what will happen. Since the situation of the Chroms is much worse than any of you might dare to think, a drastic solution is needed. People all over begin to realize that something fishy has been going on. People feel cheated. Their anger could erupt anytime and that would surely prove fatal to the Chroms. Naturally, the Chroms must survive and succeed in converting this world. Therefore, a very strong and sophisticated leader is needed. That leader is going to be you..."

"See, I knew it!"

"Please, let me finish! That leader is going to be you in an enhanced capacity. I'll be Diavo. I'll assume your identity."

"Why would that be necessary? We've been doing a fine job, haven't we?"

"Don't think too highly of yourselves! Yes, you and your people do have some special talents, no doubt. However, don't believe for a second that you could've got this far all on your own!"

"You belittle us. Besides, there're quite a few non-Chroms who appreciate what we've done for this world."

"Right... However, there's the non-Chrom majority ready to hang every one of you from the nearest lamppost. Are you willing to accept the accusations as well? Listen to me, Diavo! Let me handle this touchy situation! Let me take the blame and find a way to conquer the shallow minds of the masses."

"What if I disagree?"

"Sorry, you have no choice. I have already paralyzed some parts of your brain to make sure you don't react violently. I know you are a highly intelligent man so I left your reasoning ability intact."

"Wait a minute! If you take my identity, what's going to happen to me?"

"Very simple... You'll cease to exist."

"No! Please!"

"Come on, Diavo! I thought we've already settled that part. You said you don't think it makes a difference when your life ends."

"Well, if it's really about to end, then... then I guess it does make a difference."

"Don't be afraid! I won't hurt you. You won't feel a thing."

The blood ran out of Diavo's face. His whole body started shaking. "Can we just... talk a little more," he said feebly.

"What else is there to be discussed?"

"It's not that simple for me," Diavo replied. "It might be easier, though, if you told me what I can expect on the other side."

Wrix seemed to be at a loss for a few moments. "Well, you know, as long as you are on this side, you're not supposed to know what's on the other side. So, even if I knew, I couldn't reveal it to you."

"You mean you don't know!"

"Frankly, I don't."

"Oh, no... What's this? God doesn't know what's going on?"

"Well, I'm not the most supreme... entity in this Universe."

With his lips tight, Diavo nodded. He found some relief in Wrix's admission. Nevertheless, he went on pleading: "I can't possibly go yet. I have family back home."

"That's been also thought of," Wrix replied indifferently. "It's obvious that after the conference in Uronton, you will, I mean I will have to return to Murumbia. Assuming the role of a leader is one thing. The role of a husband, a father, a son is another. I don't wish to have that role. Therefore, your wife, your two sons, and the only other relative you have in Murumbia, your mother, will die tomorrow. Your house will burn down and they'll all die in the fire. Just between you and me, they won't suffer for a single moment. Their lives will be turned off before the fire begins and the flames get a chance to devour their flesh."

"You can't possibly do this!" Diavo raised his voice to protest, staring at the tiny round cabin-window with wet eyes.

"Now, please, look at me!" Wrix said firmly. "We don't have all night, we must do the transition."

"Are you going to stab me? Perhaps choke me?"

"Nothing of that sort, I assure you... Now, we have to interface!"

"Do what?"

"Just look into my eyes! Good... Stay that way for a few moments while I download all data from your circuits. Okay."

The next thing Diavo saw was the transformation of Wrix's entire body.

"Voila!" Wrix said with a smile. "How do I look? Can you see a difference between yourself... and yourself?"

"So, what's going to happen to me? I mean my body, now that you have created an artificial one."

"Wherever you're going from here, you'll do just fine without your two hundred pounds of perishable body... I guess."

"You guess... I must..."

Diavo did not get a chance to finish his sentence. A beam, emanating from the new Diavo's right eye, made him disappear.

"I can't take this any longer!" Twimko, the twenty-five years old Eslan said in a bitter voice. He emptied his glass and then waived to the waiter to bring him another drink.

"So, what are you going to do about it?" one of his friends asked somewhat cynically.

Twimko grabbed his friend's arm and leaned closer to him over the table. Looking straight into his eyes, Twimko shouted: "Listen to me, Bonza! Listen carefully because I'm going to tell you what can be done about it."

"Okay, okay, calm down, Twimko!" the two other young men rushed to Bonza's help. "Just sit back, will you!" They pulled Bonza's arm out of Twimko's hands.

"And stop shouting!" Bonza added. "I'm not deaf. Besides, I think you've had way too much wine already."

Twimko leaned back in his chair and then, shaking his head, he took a long, deep breath. "Never mind!" he shouted to the waiter. "Cancel that drink!" He leaned over the table again and looked at Bonza. "I'm sorry. I didn't mean to offend you. You are not the cause of my anger you know that very well... We grew up together, we've always been best friends... all four of us. You, Bonza, have always been careful. That has saved us from trouble many times... And you, Frosh, remember how you saved my life when we were kids? You pulled me out of that raging river. I still don't know how you did it." He turned to his third friend. "Zavu, you are the smartest. You got into school and now you are an accountant. You could mingle with the high society if you wanted to. Instead, you remained friends with us, me, the mason, Bonza, the farmer, and Frosh, the miner. You're a great guy, Zavu."

Zavu laughed briefly. "Come on, Twimko, you're embarrassing me... Instead, why don't you go on? You wanted to tell Bonza what you think we could do."

"Perhaps I had too much to drink," Twimko continued quietly, "but I know what I'm talking about. I know you, all of you, feel the same way about our situation. The Chroms are robbing us blind. Now that they own most of the shops here in Landis, they just keep raising the prices. At the same time, I haven't had a pay raise for years. That darn, old, alligator-faced Grunel makes us work harder and harder. A few days ago I had a chance to tell him how I feel about being so underpaid. Guess what his reaction was! He said I could resign anytime if I didn't like my job. You're not indispensable, Twimko. That's exactly how he said it... I hate that monster."

"Hey, you're loud again," Zavu warned him.

"Sorry, I keep loosing my cool... Anyhow, now it seems we have less and less hope. Our situation has got much worse."

"So, what can we do?" Bonza asked.

"What?" Twimko paused and called the waiter to pay for all of their drinks. "Let's go!" he said after the waiter left their table.

They stopped in front of the bar.

"I think it's time to start a rebellion," Twimko said as his friends were waiting for him to speak. "The city is already asleep and the street guards have probably gone home. Gosh! The Chroms are now so used to this peaceful exploitation, they think it can go on forever. They really think we are sheep. Well, I say they're wrong. We're going to show them how wrong they are."

"What are you planning to do?" Bonza asked.

"I'll mess up a few of their shops. Perhaps start with Grunel's office downtown. After that, I'll walk right through the shop-window of his son's clothing store... and set those fancy suits on fire. I might also pay a visit to Grubbi's food store and taste some of his expensive caviar. It's just fair that once in my life I also eat caviar, right?"

"It sure is, Twimko," Frosh replied enthusiastically. "I'll go with you. I think you're absolutely right. If we don't do something, nothing will ever change."

"How about you two," Twimko asked.

"Have I ever spoiled any of our adventures?" Zavu responded. "Of course, I'm coming."

"Me, too," Bonza, as usually, was the last to decide.

The walk to downtown did not take long.

"Grunel's Construction Company," Zavu read the sign under the street light. "That's it, isn't it?"

"That's it," Twimko replied.

A horse-drawn carriage appeared, turning from a side street.

They hid behind a thick tree trunk and waited until the carriage was at a safe distance.

"Okay, let's go!" Twimko said.

Zavu, the tallest and heaviest among them, simply broke through the wooden door with his shoulder. They ran upstairs and forced another door open. To their greatest surprise, they found Grunel behind his desk, leaning over some drawings.

Grunel looked frightened but he tried to maintain his composure.

"Twimko?" the fat, old man said in a feeble voice. "What... are you doing here?"

"Oh, man!" Twimko responded, shaking his head. "You sure should have gone home tonight!" He reached for the heavy, copper-based lamp on the desktop.

"What are you doing?" Frosh shouted, trying to hold Twimko's hand from reaching the lamp.

"We can't possibly let him live to tell that we paid him a visit," Twimko said as he grabbed the lamp.

Grunel raised his hand to cover his face. "Wait!" he yelled at Twimko. "I have something to tell you, young man."

"I don't want to hear it!" Twimko shouted. With one hand, he pushed Frosh aside, while his other hand hurled the lamp at Grunel.

The thick round metal base hit Grunel in the head. The old man turned off his chair and fell to the floor. Pieces of the broken light bulb scattered all around him.

"What a waste!" Zavu commented looking at the second light bulb that was hanging from the ceiling. "Some of us can hardly afford a single candle."

"Is he finished?" Bonza asked. He stepped closer to Grunel's body to feel his pulse.

"I certainly hope so," Twimko said with a long sigh.

"We'll just have to burn the place down," Frosh suggested.

"That should do for the night," Zavu said soberly.

Bonza stood up and looked at Twimko. "Nice throw. I doubt he'd ever go home again."

"Look!" Zavu pointed at the broken light bulb.

"See? We don't even have to set the fire."

They all stared in amazement as the still glowing spiral inflamed Grunel's woolen pullover. The flames rapidly engulfed the entire body and then spread to the furniture.

"There is nothing else we can do here," Twimko declared. "We might as well go home."

They sneaked out and disappeared into the night.

Originally, Gumiel was a farmer. He owned many acres of land in and around Uronton. After his wife died, he raised his daughter all by himself. Since he had already passed his prime years, he decided he would not get married again. He sold some of his land and hired a few masons to build a huge mansion next to his old farmhouse in the suburbs. Ordinary Urons wondered why anybody with only one child would want to live in a house that had over twenty rooms.

When his mansion was finished, Gumiel quit farming and established a private school for Chrom kids. Of course, the non-Chrom Urons had their ideas about what the kids were taught in Gumiel's school. All kinds of rumors were flying around.

Gumiel's house seemed to be an ideal site for the conference. The participants could discuss their agenda without the fear of being spied on. As always, the entire estate was off limit even to the neighbors. Gumiel's large dogs roamed around the house day and night, it would have been unwise for anyone to try climbing over the high plywood fence.

The "Second International Educational Conference" sign that had appeared over the gate of the Gumiel estate even before Krupchek and Shrolen arrived was nothing but a disguise.

Every country on the big continent was represented by at least one delegate, while there were no delegates from the two small continents.

Diavo was the last one to arrive in the morning, only a few hours before the conference was to begin. Some delegates proposed they wait another day but Diavo insisted he was absolutely not tired and the meetings should start right on schedule.

"Gentlemen," Gumiel opened the conference after lunch. "I'd like to welcome all of you. Special thanks to Diavo for not wanting to cause any delay in our timetable. We can imagine how exhausted he is after his long journey... Although, most of you have met before or at least know one another by name, I'd like to introduce everyone. Let me start with myself. My name's Gumiel. I trace my origin to the house of Zorel, one of Zumaniel's disciples." He laughed briefly. "Naturally, this will continue to be known exclusively within Chrom circles... I feel highly about having the privilege of hosting this conference. I'll do my best to make sure that all of you feel at home and that all of you are safe."

Standing at one end of the conference table, he pulled out a drawer.

"I invite all of you to open the drawers in front of you. Inside your drawer, you'll find a couple of handguns. They're the latest technology, made right here in Uronton at a secret manufacturing facility. You'll also find plenty of ammunition. So, we wouldn't be defenseless in the unlikely event of an attack by the late Rhodi's followers who went underground after Rhodi was liquidated... Okay, back to the introduction...We have an additional delegate from Uronia."

He waited for Morten to stand up briefly.

"This is Morten, a very skillful trader. It might come to you as no surprise that he's the owner of the aforementioned gun factory. He's fully committed to bringing an end to any uncertainty about our future..."

Gumiel went on presenting the delegates one after another. Finally, he looked at Diavo.

"Since he was the last to arrive, and since he comes from the last place on this planet..."

Everyone laughed.

"It's about time you cheer us up a bit," Diavo commented. He, too, was laughing.

"I guess I don't even have to introduce Diavo," continued Gumiel. "He's a great friend, and a dedicated fighter. Now that Flukk, the Head Noble down there wants all Chroms out of Murumbia within a year, his time has really come. He'll also tell us about his plan to overthrow the Nobles... Okay, have I left anyone out? I guess, not. All right, let's introduce our proposals... Oh, I almost forgot. The daily news releases for the local paper have been pre-fabricated for the next thirty days. Should our meetings last longer than thirty days, we would have to spend some time to come up with additional releases. Frankly, I doubt we'd need to go beyond the originally planned fifteen days... Please, feel free to make any suggestions at this point or ask any questions you might have." He drank some water. "Well, if there're no comments, I'd like to ask Komolchik to share his ideas with us."

One of the delegates raised his hand. "I just wanted to ask about the news releases you mentioned. What have you written in them?

Gumiel smiled slyly. "It's all about our discussions, our research. You know, improving the standard of education, trying to find ways of enabling the children of the poor to enroll and so on. Well, some of the Urons will have doubts and rumors will go around. So what..."

Just as Komolchik was ready to start his presentation, Gumiel's teenage daughter entered the room. All eyes turned to her.

"I'm sorry for the interruption," she said politely, holding up a sheet of paper in one of her hands. "This just came out of the telecom machine… It's for Diavo… from Rivertown. It's marked urgent." While talking, she stepped to Diavo and handed him the paper. "Sorry," she said again and then she left the room.

Everyone watched Diavo reading the message.

It took only a few moments for Diavo to run through the two short sentences. When he finished reading, he bit his lower lip. As he wrinkled his eyebrows, tears started rolling down his face. Finally, he dropped the sheet of paper and buried his face in his hands.

The delegates waited in silence.

When Diavo looked at his colleagues again, his face was all wet.

"They murdered my family," he said quietly in a shaky voice.

"Do you mind?" one of the delegates said reaching for the message.

Diavo handed him the paper.

The delegate read the message out loud: "Diavo, your house has burned down. All your relatives perished."

"I propose we adjourn our meeting," Gumiel said. "We'll continue when Diavo is ready."

"No, that's not necessary!" Diavo announced firmly. He continued after he wiped his tears off his face. "Such atrocity clearly indicates that we have no time to waste. Flukk and his men are eager to decimate our ranks."

Diavo's voice hardened as he went on, as if he had already put his family's death behind him. The delegates watched him in amazement.

"Perhaps Flukk decided to trick us," Diavo continued. "Perhaps his Decretum is nothing more than a worthless piece of paper. It's quite possible that what he really has in mind is entirely different from what's in his Decretum. By exile he might actually mean execution. Therefore, I insist that we stick to our agenda. I assure you I'm all right. I also assure you that I'm more determined than ever before to put an end to barbarism... For thousands of years, our people have been struggling to bring about peace, prosperity and happiness for all on this planet. The majority of the people have been pleased by how things evolved. Unfortunately, there're those who want to go backwards. We can't possibly allow these dark minds to continue having any power whatsoever. Until we carry out our plan in full and create enough wealth for everyone to enjoy, we must use all possible means to ensure progress, even if it means that we have to resort to the most drastic measures." Diavo finally took a long breath and carried his glance around.

The delegates looked mesmerized.

"Anyhow," Diavo added, "I'll wait for my turn to present my idea of handling the situation. What has happened to my family just reinforces my belief that my plan is the right course of action. Of course, some of you might have even more radical proposals."

Diavo glanced at Komolchik, giving him the signal to begin.

The young poet looked around. After a short pause, he said: "It appears to me that Diavo has something substantial to share with us. Therefore, I wouldn't mind postponing my presentation and letting him speak instead, provided all of you agreed."

"I put Komolchik's proposal to a vote," Gumiel said.

Everyone was in favor of allowing Diavo to continue.

As soon as Gumiel announced the result of the vote, Diavo took over again.

"Frankly, a quick fix isn't what we need. As we are pretty much an odd lot, we can count on continued harassment as long as we live on this planet. Now, let's ask the question... Do we want to be at the mercy of primitive rulers? The answer is undoubtedly negative. Therefore, we have to get rid of primitive rulers! Obviously, as long as we're talking about free societies, we, the Chroms, can't openly come to the fore. The new rulers, at least most of them, will have to be selected from the ranks of our Implants. I'm sure that only an insignificant percentage of the people suspect that there really is such a thing as the Implants, and that they're there to further our cause. The number of really dangerous individuals is insignificant. Once we gain full control of the media, it's only a question of the right kind of propaganda to discredit those individuals. Well, some of them may have to be liquidated. Not all of them, as some sort of opposition will be outright desirable. As soon as we're successful in ridiculing such opposition, the majority will not want to be associated with it. Again I emphasize, this will apply in free societies only. I see some of you are puzzled so let me get to the heart of my proposal."

Diavo took a sip of the water. He waited until all eyes were fixed on him again, then he went on.

"This world will have to be divided. Otherwise, we'll never gain full control of it. We'll leave one side free but the other one will have to be brought under a strict dictatorship. Well, here's my plan... After this conference is over, I'll board a ship here in Uronton, and travel down to Coralton. In Feroland, I'll secretly join the mercenaries and assume a new identity. As I understand, the camp is equipped to create fake birth certificates and other documents. At a later date, I'll take a group and, as missionaries, we'll enter Murumbia. We'll go to every village and town to talk with the folks. We'll convince them that the time to get rid of their exploiters has arrived. It should be easy to make them realize how bad their current situation is. They pay taxes to the Nobles while being exploited by the Chroms as well. Most of them also send money to Zupcan... I'm sure we'll gain the trust of many Sivors. Those joining us will help spread the revolution. The hope of a just and fair society free of exploitation will surely entice most of the simple folks. We'll promise them equality and rights. On our way to Rivertown, we'll overthrow as many local Nobles as possible."

The delegates paid close attention. Some of them took notes.

"Some of our Chrom brethren will be the first to flee Murumbia," continued Diavo. "This is absolutely essential to lend full credibility to the revolution. They'll go to Khamiria and establish a free government there. Some non-Chroms, even Nobles and their commanders will also flee to Khamiria where they'll be allowed to exist as long as they denounce their anti-Chrom attitudes. Finally, we'll overthrow Flukk's regime in Rivertown... I'll emerge as the leader of the revolution, using my real name once again, and take Flukk's place..."

"Those joining us in the early stages will be the regional leaders. Consolidating our regime should be no problem. The new elite, coming up from the ranks of miners, peasants and other simple folks, will tremendously enjoy their newly gained power and will do anything to preserve it. Thus, it won't be difficult for me to turn the newly formed system into a dictatorship. We'll create a secret police to keep everyone in check. The authority of our government will be absolute. We'll take control of everything, including the land. In the name of equality, private ownership will be banned. Everything will belong to the state, in other words, to us. Later, those of our brethren who didn't flee will be given the opportunity to denounce their past and openly side with the new government. We'll erase the word Chrom from the vocabulary and severely punish anyone ever using it again. This way, Chroms in the government will simply be revolutionaries. Naturally, the population will eventually be banned from practicing politics. Most of the government positions, especially the important ones, will be reserved for our people."

Diavo paused. "Are you with me so far?" he asked, glancing around.

Many of the delegates nodded.

"The borders will be reinforced and sealed. There'll be no official relationship with the rest of the world, including

Khamiria. Religious activities will be forbidden. We'll call our form of society ETA, short for Equality to All."

While Diavo was delivering the speech, his holographic mind simultaneously wondered what Ravar might think of his improvised plan. 'It was wise of Ravar to let technology tackle the problem,' he thought. 'My host computer and I needed very little time to generate this strategy. Ravar's organic brain would've taken days, possibly years, to come up with a similar plan.'

Using yet another one of his secondary circuitries, he took pleasure in analyzing the thoughts of the delegates. 'It's amazing, how skeptical these organic creatures are. Still, they present themselves as being fully confident and determined.'

His main circuitry remained occupied with the speech.

As it became clearer and clearer that Diavo's plan could plunge the entire world into a major war, some of the delegates began to move uneasily in their chairs. Of course, such apprehension did not escape the holorobot's attention; he immediately reminded his audience of the sufferings the Chroms had to endure throughout history. 'No alternative,' he suggested subliminally, re-tuning the minds of his listeners.

Diavo easily convinced everyone that after the war the Chroms would be undisputedly in control over the entire planet.

"Naturally, the struggle wouldn't end there," he asserted. "In order to make sure that our hegemony is never in any danger, we'll have to wage an all out attack all the time. A clever, hidden attack… In order to undermine the resistance of the individual, we'll have to isolate him or her from society! Alienation should be one of the characteristics of future lifestyles. We'll have to make sure that no real bonds or alliances are formed, other than our own, of course. Today, marriage among the non-Chroms is a dangerously strong relationship. Women will have to be liberated from such bondage!"

"If we succeed creating the modern woman who is no longer dependent on her man, our position can be much more secure. We might also try to drive a wedge between parents and children. A gap between the generations will also make our lives easier."

Diavo went on and on, describing ways to keep the masses powerless.

"Consumption, consumption, consumption... This should really be an area of focus. We'll have to industrialize this world beyond imagination... I'd like to point out that the future is uncharted territory. Adapting my plan means taking on an incredibly huge responsibility. We must ask ourselves whether we're prepared to accept such enormous challenge. We'd have to work very hard, indeed. I don't mean only the twelve of us present but all Chroms worldwide, including those of us with hidden identities. We'd have to work in unity and in great secrecy. Bringing our offspring into the project would require extremely careful preparation. A lot of dedicated and hard work, indeed."

"Is there really no alternative?" Gumiel asked.

There was silence.

"That's right my friends," Diavo raised his voice. "No alternative! Eat or get eaten. It's a choice between life and death for us. Please, raise your hand if you chose life!"

Though somewhat hesitantly, the delegates began raising their hands.

"All right," Diavo said, turning calm again. "Now, the only question remaining is whether my proposal is the one to be adopted. Please, raise your hand if you think you have a better, more comprehensive plan!"

The delegates looked at one another, most of them shrugging and shaking their heads.

Diavo waited for quite some time, and then he concluded: "I thank you for your vote of confidence. During the next days, we'll finalize my plan. I know it'll work. I know we won't fail. Thank you."

Part Three

(A few hundred years later)

Margotta parked her battery-powered car. Being extremely overweight, she had a hard time getting out of the vehicle. Finally, she managed to free herself and locked the car. Gasping for air, she struggled toward the main entrance of the Lipoconomy Clinic building, often wiping dirt off the parked cars with her jeans as the passageway was not always wide enough for her.

Each time she stopped for a short rest and glanced at the entrance, she had a panicky look on her face as if she was afraid she would never make it that far.

Finally, she entered the building.

There were several people in the waiting room, mainly females, sitting around in specially designed wide chairs. They were all grossly overweight. Some of them appeared even thicker than Margotta.

"Hey, lady, come here!" a woman was trying to get Margotta's attention.

"Oh! Tliana," Margotta responded. She waddled toward her friend and then slowly lowered herself into an unoccupied seat. "How are you doing?"

"You know... Just the same old..." Tliana answered. "How are things going with you?"

"Shit!" Margotta said after making a grimace. "I hate being alive. I really do."

"Yeah, I know what you mean, Margotta. Believe me, I feel the same. It's just getting worse and worse every day."

"I especially hate this clinic," Margotta continued indignantly. "I now have to come here once every few days to have the excess fat extracted from my body. The last time I even complained to the medic about this. Guess what he said... He had the guts to tell me that I should be appreciative because this way my income is higher. Of course, he doesn't seem to realize that in order to produce more fat for them I have to eat more, too."

"You know what I just heard a couple of days ago?" Tliana said as she leaned closer while lowering her voice. "It's some kind of a new talk radio station... A man called in and asked if it's true that the Lipoclinics wouldn't be in business today if overweight people hadn't been treated with hormones generations ago."

"Yeah... I'm not surprised... No one cares. We all know that our fat is used to produce soap, even the clinics don't deny it anymore. Those nicely wrapped tiny cubes in the shop-windows of Mitopolis are sold for a fortune. Some people wouldn't even consider using the old-fashioned chemicals anymore. They need real bio-soap. Well, what's new? Just look at blood consumption! It's skyrocketed lately."

"Funny, isn't it? With blood they went the other way. The most important element of life doesn't have to be bio. Who understands?"

"I bet artificial blood does a lot of damage. You can see that those needing transfusions spend more time inside the clinics than outside. Some get transfusions daily. What kind of life is that? They're probably worse off than we are."

"Right... At least we get paid."

"Frankly, I'd rather have a job if I could, and make more money. What we get here is hardly enough to get by. I keep dreaming about saving enough for a trip."

"You're crazy! What trip? We have a hard time getting around in our immediate neighborhood. Can you imagine leaving town?"

"Not really... It seems like we're stuck here, aren't we?"

"Anyway, if I could get rid of all my excess weight and lead a normal life, first thing I'd do find myself a boyfriend."

Margotta waved her hand.

"What?" Tliana raised her voice. "Wouldn't you like to make real love at least once in your lifetime? At least to see what it feels like..."

"Just keep dreaming, baby... Can't you see? Hardly anybody makes real love anymore."

One of the doors opened and a nurse called Tlina's name.

"Okay, baby," Margotta said. "It's your turn. I'll be in lab number four when you're through but don't wait for me. I'll call you when I get home."

"Welcome aboard," said the tour guide, a woman in her early adult years. "My name is Troby, I'll be your assistant today... As you can see we have just left the city limits of Mitopolis. Before anything else, I'd like to apologize on behalf of Triliom Tours. Earlier, I heard one of you commenting on the age of our bus. Well, unfortunately, all of our new, solar-engine vehicles had been reserved by Zupcan Tours."

"I thought it was now against government regulations to use these gasoline powered buses," a man sitting in the front row said.

"In a way, it is, indeed," replied the tour guide. "However, we do have a special permit that allows us to use them for destinations that are within a hundred-mile radius. The Triliom Caves are less than seventy miles from Mitopolis. Of course, I'm not suggesting that it's a great idea to use these ancient vehicles. Our planet is already in pretty bad shape. We don't need to add more to the pollution. As I understand, these buses will go to the junkyard as soon as the gasoline reserve of our company runs out in the not too distant future. Since hardly any more oil can be found in the ground, the government keeps whatever gasoline gets produced."

"What for?" the man interrupted.

"Don't ask me," answered the woman. "I'm not an expert on what the government does... Anyhow, we'll be arriving at the caves in less than an hour. As you probably know, the entrance to the caves is at three thousand feet above sea level. Unfortunately, this means that we'll have to use our oxygen masks as soon as we leave the bus... For those of you who aren't up to date on the latest data regarding the condition of our planet, I'd like to read the most recently released government memo that deals with the use of oxygen masks." She opened a folder and found the page she needed. "Okay... According to the Research Bureau, the percentage of oxygen in the air we breathe is now down to thirteen. This is the value measured at sea level. Obviously, the percentage is even lower in the cities, especially in those at higher altitudes... People living in places that are significantly above sea level more or less got used to the gradual decrease of oxygen content but those visiting at higher altitudes can suffer permanent brain damage without using oxygen masks."

"Great! Just great!" the man in the front row commented angrily. "A fantastic inheritance, isn't it? If you ask me, our ancestors were selfish idiots. They cared only about their own comfort and financial well being. They didn't give a damn about future generations."

"Please!" the tour guide tried to quite him.

"Let him talk!" several of the passengers demanded.

"Do you mind?" the man asked the tour guide, reaching for her microphone.

Hesitantly, the woman handed it to him.

"Hi group." The man turned to face the passengers. "My name is Rolli. I'm an environmentalist. I work for the government. Let me tell you something though... Much of the results of our research get ignored... I'm not saying that the government could reverse the process and save the planet if our advice received enough attention. Unfortunately, it's too late for that. We passed the point of no return a long time ago. Still, it would be possible to improve conditions and slow down the deterioration... You see, there're facts that aren't even published anymore for the sake of avoiding mass hysteria." The speaker paused briefly and then he went on. "Well, how would you feel if I told you that what limited vegetation we still have is being tremendously overworked. The quantity of the carbon dioxide we produce is now much more than what the dying grass and the remaining forests could assimilate and turn into oxygen. The balance will continue to get worse and worse until there will be not enough oxygen left to live on."

"Will that happen in our lifetime?" one of the passengers asked.

"Our lifetime... Our grandchildren's lifetime... What's the difference? Our species won't survive. That's the point. So, where is our intelligence? Buried deep under the huge piles of waste I'm afraid... I'm sorry. I didn't mean to spoil our vacation. I just can't keep quite in the face of the selfishness and ignorance that have dominated our lives ever since the steam engine was invented, perhaps longer."

"Well, is there anything we can do about the past?" the tour guide asked after the man handed the microphone back to her. "Probably nothing... However, we'll use our protective masks when we arrive. They're at the back of the bus in that large box next to the exit door. The oxygen tank is compact. You'll be carrying it on your back. Hopefully, it won't interfere too much with our enjoyment. I guarantee you'll be amazed once we're inside those caves... In the meantime, sodas and juices are available from the coolers. Just press the button in the wall next to the item you desire. A slot will pop open and the can will fall out. Same as in the solar buses... So, enjoy the ride and let me know if I can be of any assistance to you."

The tour guide, exhausted from speaking, took her seat beside the driver. She looked out the window in spite of her reluctance to do that. She had promised herself many times that during the ride she would keep her attention focused on the group.

'Incredible,' she thought as her eyes scanned the field along the road.

What she saw was one of the ugliest sights on the planet. Debris covering the field as far as her eyes could see. Aluminum cans, plastic bags and boxes, broken frames of televisions, rusting bicycles, hollow computer monitors, keyboards, speakers, and so on. Here and there the tops of corroding automobiles bulged out.

She sighed. 'Another very depressing day...'

Some of the passengers dozed off. Others put on their headsets and listened to music. Not counting the driver and the tour guide, there were two dozen people in the bus, half of them males, the other half females. They were all married couples. The men appeared to be of a variety of ethnic origins, some of them with darker skin, descendants of folks from the south hemisphere. The women all looked alike. They were clones, born in artificial wombs of the same laboratory. The trip was sort of a class reunion for them.

The men were mainly in their mid-life years with a few of them approaching old age. The women were very young.

One of the clones whose husband had fallen asleep moved to the last row of the unoccupied seats at the back of the bus. A little later, another one joined her.

"Hi, Miletta," one whispered to the other with excitement.

"Hello Luciena. It's so nice to see you," Miletta whispered back with the same level of excitement in her voice. "I'm glad we've been able to work out all the details through the e-thought channel. I'm ready. What about you?"

"I'm as ready as you are… So, let's see… Our clothes are exactly the same… Can you see any difference in our hair style?"

"No, I can't."

"I can't, either… I think we can go ahead with our plan and safely trade places… Again, the most important you will have to remember is that Grog always wants oral sex before he penetrates you. The other thing is that whatever you cook for him must have cayenne pepper in it."

"All right… As for Slover, he always wants breakfast in bed. Never sex on Zumdays… Oh, and don't forget! He lets you look at his bank statements."

"If we won't like it, I'm sure we'll find a way to change again."

"It's all the same anyway. We are slaves and that's what we'll always be… I know others who traded places before. It worked… At least we can have this much adventure in our lives."

"Okay, let's do it then! Let's do it right now! I'll just go to your husband and you'll go to mine."

They hugged and kissed.

"I wish I could stay with you forever," Miletta whispered.

"I would like the same, Miletta. When I'm with you, I feel true love. It's a feeling I have never experienced with my husband."

"I suggest we avoid talking again on this excursion."

"I agree... I'll call you in a couple of days... If we were allowed to leave the house alone, I'd come and visit you. Wouldn't that be nice?"

"Perhaps one day, Luciena... Perhaps one day... Okay, ready?"

"Ready... See you Miletta."

"Another glass of wine," Tiguri asked his guest.

"Maybe later," Furgoy said. "I seldom drink. When I do I usually stop after two glasses. So, for today, I've already reached my limit. Although I must admit your wine tastes much better than most of what I've had so far. So, indeed, I might just have another glass later."

"It's home made, my friend. You see that vineyard down there?" Tiguri looked toward the end of the orchard that began right under his balcony. "There, behind the trees... I cultivate it. It produces about fifty gallons of quality wine for me. It's enough from one harvest to another."

"You live in a beautiful place. The view from up here is breathtaking. You can see just about everything from the top of this hill and the city is only a few miles away."

"Can you see that shiny glass building down there? The one on the river bank... That's where I do my research."

"Really... It's not even half way to the city. I bet you can get to work in a matter of minutes."

"When the weather is as nice as it is today, I walk. Otherwise, I take the bus."

"You must have paid a fortune for this house. How could you afford it?"

"I inherited it from my relatives."

Furgoy sighed. "I didn't get this lucky," he said while gazing into the distance. "After we graduated... When was that exactly?"

"Hey, it's twenty years already. Can you believe it? We graduated twenty years ago and we haven't seen each others since."

"Time does fly... Anyway, I ended up doing some digging at the ruins around Coralton. That's where the leaders sent me. They knew about my anti-government feelings. They knew I sympathized with the underground movement so they sent me as far as they could. Anyhow, I married a girl I met down there. Being a native of the south, she was in pretty good health. She didn't even have any of the allergies we suffer from so much up here. Nevertheless, we lost our first child due to miscarriage. The same happened with her second and third pregnancy. During her fourth pregnancy, she died of a rear disease. Some suspected she was a victim of radiation emanating from the desert where so much of the nuclear waste is buried. Since her death, many other women suffered the same faith... I just couldn't stay down south anymore. I sent my resignation to the government and moved to Gorinton. I had some money saved so I bought myself some land not far from the city. That's where I live. Now, listen to this! One day, to my greatest surprise, I ran into Yolan in Gorinton."

"No kidding! Do you mean our Chemistry major, philosopher, beautiful blonde? Your secret love, if I remember well."

"That's right... Well, she had married a pilot from Gorinton. That's how she ended up living there. A few years after the wedding, her husband died in a crash. Not much later, her only daughter died of cancer at the age of five. I met her shortly after her daughter's death. She was devastated, as you can imagine... You probably remember how much she didn't care about me at the Uni. Well, I was really shocked by how glad she was to see me."

"Have you teamed up?"

"We've been going out... We now have at least one thing in common."

"What's that?"

"We both strongly believe that this government has been in power way too long. We are also convinced that it is still not late to save our world from destruction... If we could only find a way to overthrow this so-called freely elected, democratic regime..."

Tiguri shook his head.

"I guess you see it differently," continued Furgoy quietly.

"Unfortunately, I do... We do a lot of gene research at the lab. Results show that the immune system of our species got damaged beyond repair. Due partly to mutation, most of the medications are no longer effective. In fact, a lot of the synthesized drugs now attack our genes instead of attacking the viruses and bacteria. We now have a planet full of weak and sick people. Babies are born with chronic diseases. It's a tragedy of great proportion... Just look at our skin! Without the medications in our shampoos we would all look like lepers. A few generations ago some clever businessmen figured out a way to make all of us dependent on the medicated shampoo. Here's the result."

"How did they do that?"

"How... It's very simple. The same way they did it with everything else. They began feeding us a lot of synthetic stuff. These chemicals were used as additives not only in shampoo but also in our food, in our drinks. People were led to believe that the additives would enhance the body's natural defenses. What happened instead was that our system got confused. The body stopped producing the hormones and other essential substances. Now, withdrawing the additives would mean quick death. The biggest problem is that this dependence is carried on into our newborn. Everyone is sick. Can you show me at least one single individual who's not afflicted by some sort of degenerative disease?"

"The clones, perhaps..."

"Oh, yeah, the clones... How can I neglect mentioning the clones? The promise of the future as one of those idiots in Zupcan calls them."

"I see you don't think too highly of them."

"Of whom... The clones or the idiots?"

They laughed.

"Neither, I guess," Furgoy said reaching for his glass. "You know what? I'm ready for some more wine."

"I'm glad you are... Hey, you remember how health conscious I used to be? I'm sure you do... Now you must be wondering whatever happened to me."

"You're right. I'm a bit puzzled, seeing your drinking habit."

"Well, I don't know about you, Furgoy, but as the years went by and as I saw the death of my wife, then the death of my only son, I started thinking. I concluded that our brains have their limitations. The universal puzzle of why we are here and where we might go once this is all over remains unsolved. Some of us began to ask these questions at the age of six. You can grow to be a hundred years old and you'll be still asking the same questions. Therefore, I decided that I don't care how many millions of my brain cells die when I drink a glass of wine. I enjoy it and that's what matters most."

Furgoy slowly sipped his wine.

"Now, what's this sudden silence?" Tiguri asked after a while. Leaning closer to Furgoy, he raised his hand. "Wait a minute! I know you... You have something on your mind but you're not sure whether you should say it or not. Right..."

Furgoy smiled. "You're absolutely right, my friend."

"Hey, just go ahead and say it! No need to hesitate."

"I was just wondering whether you might be interested in joining... us."

"Us... Who would that be?"

"The movement, of course... The opposition to the government... You could do great things for our species down there in that research lab."

"In a way, I'm already a member of your movement, my dear friend," Tiguri answered with certain pride in his voice. After emptying his glass, he leaned very close to Furgoy. "I'll let you in on a couple of secrets," he continued lowering his voice a bit. "I have single-handedly sabotaged quite a few drug research experiments. When I realized what further harm those new drugs could mean, I managed to fool the research team and thus cause the experiments to fail."

Furgoy stood up and stepped to the rail at the edge of the balcony. He glanced down toward a van parked right behind his own car.

"Don't you want to hear the rest?" Tiguri asked somewhat offended.

Furgoy raised one of his hands above his head, and then he turned back to Tiguri. "Oh, I'm listening," he said. "Just go on… I'm really listening. It's all very interesting."

Tiguri narrowed his eyes and bit his lower lip. Pointing his finger at Furgoy, he mumbled: "Now, wait a second, will you?" Suddenly, he turned his head and looked inside his living room. "What's that awful noise out there? Is someone breaking in through the front door? I bet that's what…" He did not finish his sentence. He saw two men in military fatigue taking up positions inside his living room, pointing their laser handguns at him. Again, he looked at Furgoy.

"Sorry, my old friend," Furgoy said as he leaned against the rail. "You've just got busted."

"You... You!" Tiguri could hardly speak. "You set me up!"

"Someone had to do it… Again, I'm sorry it had to be me." Furgoy reached into his pocket and pulled out a slim recording device. "Your confession is now in the chips. These men will take you now, go and follow them. Don't resist and don't ask any questions! Just go! Oh, by the way, thanks for your hospitality."

"Can I still trust you?" Dobias asked his girlfriend after they stopped kissing.

"What got into you?" Liazel snapped back at Dobias.

The young man, caressing the girl's fingers, looked out the window.

"Come on, Dobias! What's bothering you? Talk to me!"

"I'm not even sure," he said. "It just feels that everyone here in Landis has gone dumb. Everyone seems to be accepting this... this new-world order. This one planet one government rubbish... As if resistance was absolutely futile... So, I'm just wondering whether you're still with me."

"Of course, I am, you silly boy," replied the girl. "However, I think you worry a bit too much lately. I hope this won't affect your studies. As you know the finals are just around the corner. You still intend to graduate, I hope."

"You know, I've been wondering about that."

"Just what do you mean? You have a job lined up... Finally, you can get out of this dorm and rent yourself a decent place."

Dobias raised both his hands. "Just hold on a moment, will you? Please!" He leaned his back against the headrest and put both his feet up on the sofa. "Quite frankly, I'm not sure I want to be a servant of this Mafia running the affairs of our world."

"Mafia... That's a bit too strong, don't you think? Besides, you should watch your words. Remember? You told me that listening devices are all over these days. You don't want to get in trouble, I hope."

"I doubt they bug dorm rooms. Otherwise, they would have picked me up long time ago... Liazel, I don't mean to scare you but our future isn't as bright as they would like us to believe. Just think about it for a moment... The Eslans are giving up hope of ever becoming independent again. That's a great disaster... Our ancestors fought hard against the Chroms in the war..."

"Against the Chroms?" the girl interrupted him. "The war happened between the Eslans and the Sivors."

"Yeah-yeah, sure... According to the history books... The problem is that history has been re-written several times. Besides, these days you don't even hear the word Chrom anymore. It has become a taboo. I should say the Chroms have managed to make it a taboo. Say anything negative about them and you're in trouble. Once you're branded anti-Chrom you can forget social acceptance."

"Well, it's the same with everything else, isn't it? You can't be anti-gay, you can't be anti-moron, just can't be anti-anything."

"That's exactly my point. You can't be negative about anything that's negative. Mind-boggling! No wonder so many decent people commit suicide. The situation is hopeless... Just look at where we are today! It's basically a one-man dictatorship. Chancellor Wrixman dictates the terms of our democracy from Zupcan. His cabinet members are spineless hypocrites. Of course, they get re-elected over and over again simply because the majority of the people depend on them. Half the population is terribly sick. The government subsidizes their medicine. They'd never take the chance of voting for another government. These votes alone will keep these bastards in power forever. Anyhow, talking about the war... I know it from my father and he knew it from his father. Luckily, my folks never yielded to intimidation. They never hesitated to pass on the truth to their children. I know it's hard to believe but we do live in a world controlled by the Chroms. They've been manipulating our lives since the beginning of time."

"So what... Is there anything you can do about it?"

"The problem is that everyone has the exact same attitude. In the meantime, our world gets destroyed. Anywhere you look you find the works of negative forces."

"Your imagination is wild, Dobias. You see enemies where there's none."

"What imagination? Okay, here are a couple of facts. Look at this building, for example. It stinks! You can smell it from a hundred feet. It's some kind of chemical smell. I bet they used a certain type of plastic during construction that gradually degrades over time and generates this toxic fume. I don't know how it might affect our health in the long run but I'm sure it's not harmless. Go to any newly built public buildings, schools, libraries, you name it. They all have this same smell. What's this if not an attempt at undermining our health even more?"

"How come I can't smell anything?"

"Right on! I just asked one of the clerks at the library how she can put up with that really strong smell. You know what she said? What smell? That's what she said. Millions can't even smell anymore. People's senses are destroyed as well. Wait! I'll give you another example. There's so much talk about sleep deprivation these days. The number of accidents caused by people falling asleep while driving is at an all time high. People just can't sleep during the night anymore. Sleeping pills are being pushed like never before. The root of the problem, of course, is shunned. There're at least a dozen new laws favoring the noisemakers. In fact, I think there're regulations that outright dictate the creation of noise. Just look at the railroad alongside this campus! Naturally, freight trains are permitted to run only during the night. They are, of course, required to blow their very loud horns at every intersection. Now, I really don't get this one. Those road barriers automatically come down to stop traffic way before the trains reach the intersections. Why blow the horn then? Well, the law says it's for the protection of pedestrians who might accidentally wander onto the railway tracks. Excuse me! How many people trains have hit during the last few years in the entire city? Two... One was a drug addict and the other one was an alcoholic. There're quite a few trains going by here. Their loud horns wake me up at least a dozen times each night. Thousands and thousands of people who live along the rail lines must go through the same darn punishment just in this city alone."

Liazel frowned.

"Look!" said the girl. "I know you're probably right. However, you shouldn't work yourself up like this. It's just not worth it."

"Come on, Liazel, survival is at stake."

"I'm sure we'll get through life just fine. We live in a modern world that has its disadvantages. We can adopt. Besides, where is it written that life must go on here forever? Okay, life could be undoubtedly safer if we still lived in the Stone Age. Although, a simple snake bite could finish you off… I think we're well compensated for whatever isn't right in our lives today. We live in an exciting world of comfort and knowledge. True, our planet is badly polluted. Yes, life will be extinct here in the not too distant future… Let's suppose for a moment that we haven't created technology. Let's suppose our environment was as clean as it could possibly be. Then what? How many more generations would follow? A hundred, a million… What's the difference? The sun will eventually cool off and there will be frozen darkness here."

Dobias' eyes opened wide.

"Gosh!" he said. "I've never heard you speaking like this… I just can't believe you're saying all this. You actually sound like… like a Chrom."

Liazel stood up.

"Guess what buddy! I'm a Chrom… and I'm fed up with your conspiracy theories. Besides, I've developed a bad condition of genital warts recently… and since I know how paranoid you are when it comes to illnesses I really doubt you'd want to make love to me ever again. As for my mouth herpes, which I should've told you about some time ago, I've been treating it very effectively with anti-bacterial and anti-viral ointments. I doubt I have infected you. You'll be all right. You'll probably live forever. Good bye and don't bother calling."

The girl stormed out.

Dobias stood up and walked to the fridge to fetch an apple. After the first bite, he sat back on his sofa. Looking at the white clouds through his window, he mumbled:

"Well, what's new? Aren't they all over?"

'WELCOME TO THE TENTH CONFERENCE OF THE UNITED PLANET', the banner read above the entrance to the Big Hall in the Zupcan government complex. The hall inside was decorated with lots of flowers. There were nine seats facing the podium. Seven of the seats were already occupied. A huge picture of a stocky man photographed in military uniform hung on the wall right above the podium.

The delegates, one from each of the jurisdictions that covered the territories of the former states, chatted quietly:

"Is it true that the delegate from the former Khamiria lost his life on her way to the conference?"

"That's the rumor I've heard as well."

"It's strange, to say the least. Two major accidents in one day... The representative from the North Continent could not make it, either. His plane plunged into the ocean only hours before the second crash happened."

"I don't know what to think, frankly... Terrorist acts of the opposition?"

"What opposition? Terrorists were weeded out long time ago."

"Perhaps my worst fear is coming true. I don't really want to be specific at this time, so let me just say that I'm a bit worried about recent activities of our Central Police Force."

"Well, both of the missing delegates have publicly voiced their concerns regarding the environment."

"I didn't want to admit it even to myself but lately I've been wondering whether we're cutting the tree under ourselves."

"Now that you dare to mention this, I would also like to say that we might just have gone too far modernizing our world."

"I'll be blunt. It's no use beating about the bush... We are responsible for all that's going on and no one else can put an end to destroying our environment."

"I really think it's too late."

"Better late than never... I strongly believe we should do something before it's really too late."

The door behind the podium opened and a heavily armed policeman entered.

"Ladies and gentlemen!" the policeman said. "It is my honor and privilege to announce the arrival of Chancellor Wrixman."

The delegates rose to their feet and customarily greeted their superior with a round of applause. The Chancellor raised both his hands, waved to the delegates and then he took his seat.

The policeman positioned himself standing behind the podium.

"Please, take your seats," the Chancellor said in a somewhat harsh and metallic voice. "Thank you all for coming. Unfortunately, as probably all of you know it by now, two of the delegates didn't make it... Well, here's another proof that our world is still far from being perfect, as accidents still happen. I suggest we all stand for a moment of silence."

The seven delegates rose in unison and then seated themselves again when the Chancellor instructed them to do so.

"I've already sent two skilled politicians to take over the offices of the deceased. They're well updated as both of them come from my inner circle of friends. Be assured that they're capable and skillful experts. They'll be able to assume their new duties without any problem."

Wrixman took a sip of water.

"In all fairness, I wouldn't be surprised if some or even all of you had some concerns lurking in the back of your minds. After all, we do hear all sorts of rumors about the condition of our planet. This issue seems to be more pressing now than it did at the time of our last conference, therefore, I'd like to spend a fair amount of time discussing it. I personally believe that the problem is blown out of proportion. I'll be able to prove that no irreversible damage has been done. I'd like to give you my word that when we arrive at a really critical point in our development, I'll do whatever it takes to compensate for any negative side effects. We have the knowledge and the technology to implement even drastic measures if necessary."

He softened his voice as he continued.

"My dear friends… All of you know how hard we have worked to get to where we are today. We have come a long way, no questions about that. Thanks to you, and those before you, today we can all sleep without fear. No Korvags or Eslans will brake through your doors to kill you and your family. This is a hard fought achievement. Just imagine the insecurity and the worry many of your ancestors had to live with… or die with. You don't want those days to return, do you?"

The Chancellor looked at the delegates for a while as they were shaking their heads.

"Please, repeat after me… We don't want those days to return!"

The delegates did exactly as Wrixman asked them to do.

"Okay, I'm glad your faith is reinforced," continued the Chancellor. "I'm sure even the higher powers are pleased… whatever they are and wherever they might be. We have no way of knowing just how exactly these powers function. One thing I do believe is that when a person's faith is shaken, all sorts of things can happen to him or her. Yes, even plane crashes. So please keep the faith! It's in your own best interest."

By the time the Chancellor was ready to move on to the next point on the agenda, the delegates no longer had doubts about whether or not to continue in the direction their forefathers laid down generations earlier.

Ravar's fingers were busily tapping the control panel when Lella woke up from her brief nap. The woman stretched, and then she cheerfully yelled at the man:

"Hey dude, what's the scoop?"

Ravar stared at her with a puzzled look on his face. "What have you just said?" he asked.

Lella smiled broadly. "I said hey dude, what's the scoop?"

Ravar frowned. "Excuse me for not appreciating your humor but... I simply have no idea of what you're saying."

"Oh, never mind." The woman waved. Her smile faded. "I've just had this crazy dream... I was chatting with some youngsters in the streets of New York."

"New York?" Ravar was shaking his head. "And just where would that be?"

Lella pressed a fingertip against her forehead while searching her memory. "Hey, I've got it!" she exclaimed, snapping her fingers. "Do you remember that tiny planet we passed I would say a couple of days before the Arrow broke down? You know, the one that was transmitting all sorts of messages into space, hoping to make contact with alien life forms."

"Wait!" Ravar raised his index finger. "I think I know what you're talking about. We even observed some of their local transmissions."

"Right... They called their planet Earth, and they referred to themselves as humans."

"Yes, I remember. The Captain jokingly commented that we might as well call ourselves humans, since they looked just like we do."

"Except they were so much shorter... They looked exactly like these creatures down here on Y2122. Maybe they're cousins or something."

"You mean maybe they were... Those humans might have all killed each others by now."

"That's true... Anyway, I did enjoy watching some of their television transmissions we recorded. I'm pretty sure that's where my dream comes from... So, what's the scoop?"

Ravar laughed. "Okay, you probably checked the universal translator. Don't you think it would be fair if you finally enlightened me?"

"What's happening? That's what it means or something like that."

"I see," Ravar responded, still shaking his head.

"So, is there anything new?"

"I can't believe this," the man said. "Come and take a look. Wow! Chancellor Wrixman... Head of the United Planet..."

"Let me see!"

"Yeah, just look! Our holorobot took total control of the situation down there. Very creative, I must add... He has liquidated half a dozen people and assumed their identities in order to be re-elected as Chancellor half a dozen times in a row."

"Is he way out of line?"

"I haven't fully assessed the damage yet... Well, there's still life on the planet, and that's a good sign."

The woman kissed Ravar's earlobe.

"Perhaps we've spent a bit too much time making love?" she whispered giggling.

Ravar kept on musing.

"It's really astonishing," he said. "How did he come up with the idea of creating a generation of clones? I've never programmed him to do that. In fact, the idea has never even occurred to me... I only gave him more freedom to improvise."

"Well, is it a good or a bad idea?"

"It's very clever to say the least... The problem is that clones may not present a solution."

"Do you mean a solution for us?"

"That's what I mean."

"Why not..."

"You see, my dear Lella, if the original inhabitants die due to lack of oxygen, the clones die with them."

"Ravar, what are you saying?"

"I'm afraid you're right... We did get preoccupied with ourselves a bit too long and neglected to pay enough attention to Wrix... Now, we'll definitely have to keep an eye on him. I believe the situation is not totally hopeless. It may be possible to save at least the clones. The rest will probably not survive."

"Fewer problems to tackle when we get down there," commented Lella.

"That's one way of looking at it," Ravar replied. "Anyhow, go ahead and have breakfast. I'll start working on modifying Wrix's programming. I want to find a working solution as quickly as possible. If everything goes well, we could be on the surface before this day is over."

"Wouldn't that be nice?" Lella said as she stepped over to the food processor. She briefly studied the menu, and then she ordered two power biscuits and two portions of fruit juice. After she had placed the two thin-walled titanium cups under the dispensers, she waited for the juice to start flowing but nothing happened. She pressed the "ready" button again... Nothing again... "What's happening?" she asked loudly so that Ravar could hear her well.

Ravar's fingers were busy working the touch-screen panel of the central computer. When he looked up, Lella saw the concern on his face.

"It's not working, right?" Ravar said.

The woman nodded.

"I don't understand." Ravar was rubbing his forehead. "Our main terminal has just gone off-line."

"What's that supposed to mean?"

"It means that we lost power... I won't even be able to send data to Wrix."

"What's happening?" Lella asked anxiously.

"Well, something that's not supposed to be happening," Ravar replied, trying to be calm. "It could be a faulty relay... I'll have to check each and every one of them manually."

"What if you can't correct the problem?"

"You know darn well what happens when the air doesn't get circulated and re-oxygenated."

"All right," Lella sounded upset. "You don't need to get irritated with me. I didn't cause this, did I?"

Ravar stepped over to Lella and hugged her. "I'm so sorry," he apologized. "I didn't mean to lash out at you. You know that." He kissed her. "I'll find the problem, I promise. We won't get..."

He could not finish his sentence. A loud warning signal sounded and the main viewer came alive. A moment later, the face of their Captain appeared on the screen.

"Good day to you Ravar and Lella," the Captain greeted them. "I hope both of you are well."

The astonished couple stared at the Captain's face.

"We're on our way to your coordinates," the Captain continued. "As a precaution, we've sent a command to disable the central processor so the Shuttle couldn't be taken from its current position. Don't even try to override the command, it won't work. We're monitoring your life support system. There's no need to worry about decreasing oxygen level inside the Shuttle. We are already at the outskirts of the solar system and should reach you shortly."

"Captain," Ravar took advantage of the pause. "Have you... been able to repair the Arrow?"

"That's correct, Ravar," came the Captain's response. "Very soon after you two lost faith and escaped, we were able to bring the main terminal on-line. Then it was only a question of time to do all the necessary repairs. Finding you two was another matter altogether. The Shuttle appears to have sustained relatively heavy damage when you exited the Arrow. The external communication relays must have broken off. Our long-range scanners were unable to detect you. What we found instead was traces of frozen oxygen. We concluded that the gas leaked from the Shuttle. We followed the trail until we could see you. You can consider yourselves extremely lucky to be alive."

"Am I glad to see you, Captain," Lella said with a big sigh of relief. "We thought we've found a planet for ourselves... We've tried to make some changes on the surface..."

"Lella!" Ravar interrupted her angrily.

"Come on Ravar!" Lella snapped back at him. "It's time to admit that we made a mistake. We were cowards."

"Speak for yourself!" Ravar shouted.

"If I do that I must also say that it was not I who brought this planet to the verge of catastrophe. You promised you knew how to bring about positive changes in the attitudes of the inhabitants. Instead, you've created conditions that render the surface almost inhabitable. You and your genius holorobot…"

"All right, all right!" they heard the Captain's firm voice. "You two are young enough to be forgiven for some foolish mistakes… We are now within transporting range. As soon as my chief of security and I beam over to the Shuttle, life support will be re-established. Be prepared to explain what exactly has expired since we lost you. We'll make an attempt to correct whatever damage you've caused and then we'll be on our way home. Oh, by the way, Lella, your mother sends you her best wishes. She's been worried about you. She also informed us that the Arrow 2 was to be launched and sent to our rescue. Fortunately, they received our message just in time."

The Captain left her command post and a bit later, she and Trog materialized aboard the Shuttle.

"Relax!" the Captain gave the order as soon as she was able to move.

Ravar was sitting in his seat burying his face into one of his hands.

Lella's wide-open eyes were fixed on Trog's hand resting on his laser handgun.

"No need to despair," the Captain said cheerfully. "Ravar, you have every reason to celebrate. After all, you'll live... You may have to answer a few questions when we get home… Please, understand that I'd have to report this incident even if the Shuttle didn't get damaged. Now, let's see the logs! I need to have a clear picture before making any decisions about what to do here."

Ravar looked at the Captain. His face reflected his internal struggle. It took him some time before he was able to speak. His words surprised everyone, probably himself as well.

"I'm sorry," he began slowly. "I'm really very sorry... I'm afraid my ego is bigger than the entire Universe." He looked at Lella. "I'm sorry for talking you into this... this adventure. It could've ended much worse, indeed... I was selfish. I needed someone to look up to me. You've been a perfect support for me, Lella... Aside from what happened, I really love you."

Lella stared at her fingernails and was unable to say anything.

Ravar turned back to the Captain.

"I should've known better," he continued. "I thought I was already prepared to meet every possible challenge. I must admit I overestimated my own potential." He paused, and then, very quietly, he added. "I can't be forgiven for what I've done to the population of this planet. I've accelerated their aging process so badly..."

"Let's see!" the Captain interrupted.

After Ravar took her through a summarized version of events, the Captain smiled. "You know what, Ravar?" she said. "You actually deserve to be complimented."

"What?" Ravar asked in surprise. "I mean what for?"

"I think you showed a lot of compassion here," the Captain answered. "Do you think I would've acted much differently seeing the merciless acts of those Korvags? You may have had your own goal in mind but your intervention was actually guided by your sense of justice."

"Captain," Ravar cried out in a guilty voice. "I have simply switched roles around. I took the victims and made devils out of them. That's hardly anything commendable."

"What's done is done," said the Captain. "However, the same could very well have happened without your actions. I doubt the Chroms would've put up with all that cruelty much longer. I'm sure, sooner or later they would've figured out a way to turn the balance in their own favor... Frankly, there's another way of looking at this. Without your intervention things could've turned out much worse. One side might have completely wiped out the other, just to mention one possible outcome. At least this way all of them had a chance to avoid premature death. This species, living on this insignificantly small planet, wouldn't have had a chance to accomplish much anyway."

The Captain gazed out the window for a short while.

"During my voyages through the galaxies," she went on, "I've seen a number of much more powerful civilizations completing their cycle of existence without ever amounting to anything significant as measured by the family of advanced species that search the galaxies. A civilization dies out every moment somewhere in the Universe. At least you've shown some light to the inhabitants of this planet. Their sun is burning up rapidly. Who knows how far they would've evolved on their own before their world gets frozen?"

The Captain continued in a motherly voice.

"Let me assure you, Ravar, you're not the first one to act in an unusual manner under unusual circumstances. Even the Protocol allows some room for incorrect actions provided they're not criminal in nature. This was one good lesson for you, no doubt. A real life experience, not something you could've learned in school... Now, let's see what the most appropriate way of withdrawing Wrix from the surface is. We shouldn't waste much time bringing this incident to a conclusion. After all, our loved ones are anxiously waiting for us at home."

Dobias made up his mind. He was determined to change the course of history.

After he got off the train in Zupcan, he headed straight for the headquarters of the government. The laser gun in his pocket was set to kill. He had acquired the weapon from a drug dealer who even showed him how to use it. He had told the dealer he was after a cop.

Dobias saw the first guard at the government building when he arrived at the arcade leading to one of the side entrances. The guard was looking intensely in the other direction; Dobias was able to sneak in behind him.

'Boy!' he thought 'this one was really easy.'

He had suspected he would have to get through a handful of the guards in order to get inside.

He encountered the second guard at the revolving door. That guard was trying diligently to help a woman whose long skirt got caught by one of the wings of the door. Dobias slipped through the side door without the guard ever noticing him.

The third guard at the reception desk was on the phone engaged in a heated conversation with someone. Dobias simply walked by him.

The first real challenge presented itself inside the elevator.

"May I see your ID?" the armed operator told Dobias after the door closed.

Without hesitation, Dobias delivered a powerful punch to the guard's left chest, aiming at the area of the heart. The guard passed out and collapsed.

After finding the keys on the guard's belt, Dobias locked up the elevator and slipped out on the top floor before the door closed.

'There must be at least one more guard before I get to the office of the Chancellor,' he thought. However, to his great surprise, and even greater relief, he encountered no one in the corridor.

"Well, this would be it," he whispered to himself.

He stood in front of the door and read the sign "Chancellor Wrixman" at least three times. He wanted to make no mistake.

He knocked on the door.

"It's open," the answer came from inside.

Dobias entered.

The Chancellor was sitting behind his desk.

"What can I help you?" the Chancellor asked in a friendly voice.

"I just..."

Wrixman interrupted him: "Close the door behind you, will you!"

"Oh, of course..."

After closing the door, Dobias approached Wrixman, stopping only a few steps away from his desk.

Wrixman stood up and smiled.

"Well, young man, what can I do for you?"

Dobias suddenly pulled out his weapon and pointed it at the Chancellor.

"I doubt you could do anything for me that I'd like," Dobias said in a very disgruntled manner. "You've done way too much already! It's time that your actions stop."

The young man pulled the trigger, releasing a thick beam of laser from the tip of the gun. The beam hit Wrixman in the middle of his chest. However, as soon as the beam hit, a sizable hole opened up in Wrixman's chest, allowing the beam to pass through without touching any part of the body. The hole disappeared as soon as Dobias took his finger off the trigger. Wrixman was totally unharmed.

"You might as well put that thing down," the Chancellor said jovially. "I assure you it doesn't look too good in your hands."

"What the heck!" Dobias shouted. "Can I believe my eyes?" He fired again. At this time he aimed the beam at Wrixman's head.

"Wow!" Dobias uttered when he saw Wrixman's whole head disappearing, allowing free passage to the fiery beam.

He angrily hurled the weapon at the Chancellor. It passed through his body as if flying through thin air.

"Are you done?" Wrixman said smiling. His entire body was visible once again.

Dobias stared at the Chancellor helplessly.

"I should've suspected," he mumbled. "Your guards..."

"What about them?"

"They offered no resistance!" shouted Dobias again. "I passed through to your office as if I was going to a public library. That can't be a coincidence."

"Perhaps your Gods had arranged it for you," Wrixman offered an explanation.

"My Gods..." Dobias yelled. "And just what are you? Are you the devil? Killing life on this planet? You and your predecessors! I know... I know what you've been doing. I thought it was time to stop you."

"Calm down, my friend," Wrixman said. "Calm down... As you see, you can't harm me. Well, I'm not going to hurt you, either. In fact, I'm glad to see you. You could very well be the solution here."

"The solution... What solution?"

"First, calm down... Let me know when you're able and ready to listen intelligently. After all, you strike me as a clever young man."

Dobias tried to quell the storm inside him.

Suddenly, he was curious.

"You're a very strong individual," Wrixman continued. "I hope you're fully aware of that."

"Yeah, I must be," Dobias mumbled. "Except, I have no idea of what's happening here... I must be dreaming or something..."

"You're quite awake, my friend, I guarantee you that. Furthermore, your confusion will clear up as soon as you allow me to explain the situation."

"Explain the situation?" Dobias shouted again with anger. "You mean you're going to wash your hands of all the destruction you've caused in our world?"

"Will you listen or will you not?" Wrixman sounded dead serious at this time. "If you're not able to regain your control, I'll have to destroy you as well. Please, don't force me to do that!"

Dobias' behavior changed in an instant. "I'm sorry," he uttered.

"No need to be sorry," Wrixman replied. "You're about to learn something that'll distinguish you from the rest of the inhabitants of this planet. If you're ready to listen, take that seat!"

"Your seat..."

"That's right, Dobias, take my seat!"

"How do you know my name?"

"Please, son, stop insulting me! Just sit down, make yourself comfortable, and maintain an open mind."

After a few uncertain steps, Dobias lowered himself into the Chancellor's chair. He remained on full alert holding on to the armrests with both his hands.

"Relax," Wrixman said in a comforting way. "Believe me! I have no intention of hurting you. It's time for you to realize that I could've already vaporized you if I wanted to. After all, I'm the devil as you put it. Well, my friend, that's not exactly the case... First of all, you need to know that I didn't come here on my own."

"No?" Dobias felt compelled to say.

"You also need to know that I'm not really alive... not in an organic way that is. In other words, I'm not made up of tissues like yourself."

"Then just what are you made up of?" the startled young man asked.

"Photons, ions, electrons, and so on... Don't even try to comprehend it. It's technology that your folks haven't discovered yet. Whether they ever will might actually depend on you."

Dobias kept shaking his head in disbelief.

"Hey, just stay focused, okay?" Wrixman tried a more cheerful tone.

"Okay, I'll try."

"Good... Now then, I've already told you that I was sent here. What I haven't told you is when, why and by whom... My real name is Wrix, just so that you know."

As Dobias began to pay close attention, the holorobot suddenly transferred essential knowledge of past events to the young man's brain, careful not to overload his fragile, organic circuitry. Dobias never realized that the transfer took place in a single moment. What he remembered was the Chancellor telling him the story over a certain period of time.

"So what," Dobias exploded as soon as the transfer was complete. "You could've done things differently!"

"That's true," Wrix fired back right away. "In that case, quite possibly, you would've never been born... So, take it easy! Whatever happened had to happen in order for us to be right here right now."

"Is that something you consider good?" Dobias argued. "I mean being right here right now?"

"You tell me, my young friend! Would you rather not exist?"

Dobias could not answer. He just kept blinking nervously.

"Okay, Dobias, let me continue... My assignment here is over. I am being recalled. I'd like you to consider taking my place."

"Taking your place?"

"That's exactly what I've said... You realize that my work here is unfinished. Someone will have to carry on. You more than qualify. Your Gods could not have sent a better replacement."

"What are you talking about?"

"Are you listening to me? Dobias! There's no point in refusing my offer."

The young man sprang to his feet.

"You listen to me, you photonic son of a gun!" he shouted. "You've spent a long, long time derailing the work of nature on this planet. Now, instead of making an attempt at least to correct some of the damage, you simply bail out. That's not even all! You want me to take your place and... and get lynched by the masses."

"Shut up!" Wrix yelled at Dobias. "You have no right to talk to me like that! You came here to kill me. I spared your life, didn't I? So, stop playing the smart guy! Come to your senses! I'm actually offering you power you currently can't even imagine… Now, I'll be blunt with you. I could simply force you to accept my offer. You would never even realize that you acted against your own will. However, I won't do that."

Wrix paused.

"Frankly, now that you possess the knowledge, you can't just walk out of here. Well, just imagine for a moment that you actually did walk out of here. What will you do? Will you tell the people that you had a conversation with the Chancellor? Will you tell them what you've learned here today? You would immediately be branded insane and locked up in an asylum. No one would believe your story about passing through the guards, firing at me… However, as I said, I'm not going to force you. You can freely walk out of here the same way you came. No one will ever know of your encounter as long as you keep your mouth shut. The question is whether you would be able to keep all this to yourself. I refuse to speculate about the excruciating mental torment you'd experience during the coming years… Well, as an alternative, you might ask me whether I could simply erase this experience from your memory. Of course, I could. Would you really want that to happen? Would you? Of course, not! What's life's worth without knowledge? Experience and knowledge, Dobias… These are key ingredients. So, think about it! In addition to knowledge, you'll also possess power. You won't be quite like me, of course, but you'll be powerful enough to rule this planet. You won't be able to make drastic changes without plunging your world into total chaos but you may affect the outcome in a positive way."

"What if I really decide to walk out of here?" Dobias asked. "Will I be able to resume my simple life?"

"Well, if that's what you really want… You see, you're not the only one capable of taking over from me."

"My Gods would just send someone else, right?"

"Don't be sarcastic, young man, and don't make a rushed decision, either!"

"Wait a minute!" Dobias yelled. "I'm just realizing something here... Don't tell me that I, Dobias, can simply sit in your chair and show my face to the world after you disappear! The first guard that walks in here will shoot me on the spot. At best I'll stand trial for assassinating you."

"Poor fellow..." Wrix made a face. "Do you really think I meant to leave you behind in your current appearance? Of course, I wouldn't do that. You'd look just like me. Here, take a look! Voila! There's a mirror on the wall right behind you. Look at yourself!"

Dobias could not believe his eyes.

"Incredible," he whispered as he checked out his new body. He turned back to Wrix. "Man, I look just like you... And so, now there are two of us. What about that?"

"Don't worry about that! Once I am gone, you'll be the only one left."

"One more thing though... What would happen to me? I mean to my old body."

"Not a problem... Your lifeless pile of organic debris is right there by the door, somewhat fried by the Chancellor's laser pistol. You can smell that unpleasant odor of burning flesh, can't you? You were able to outwit the guards and sneak up here but the Chancellor was alert and gunned you down. Your friends and relatives consider you a hero. However, for the time being..."

Suddenly, Dobias had his own body back and his corpse was no longer lying on the floor.

"It's up to you, Dobias," he heard Wrix's voice.

"Where are you? I can't see you anymore."

"I'll leave you alone for a while. I know how overwhelming this experience has been for you, so I'll give you time to think undisturbed. When you've made your decision, call my name. I'll return immediately."

"Wrix! Wait! Don't go anywhere! I've already decided... I'll take the job."

After getting out of the elevator, still panting, Margotta searched for her keys. Her fingers, shoving the dozens of medicine bottles out of the way, reached the bottom of her oversized purse.

"Boy, I hope I didn't lock them in my car," she mumbled to herself.

She checked the pockets of her raincoat.

"Oh, there you are!" She sighed with relief.

The entrance to Tliana's apartment was just a couple of doors down the corridor. Margotta rang the bell to let Tliana know of her arrival and then she opened the door.

'I wonder how she's doing,' Margotta thought as she entered the apartment.

Tliana was in bed with her eyes closed. She only woke up when Margotta touched her arm.

"Hey, baby!" Margotta greeted her trying to sound cheerful. "How are you doing?"

Tliana's puffed-up face was twitching as she tried to smile. "Hi, my love," she said in a feeble voice. "Thanks for coming."

"How do you feel?"

"It's over, my love... It's over," Tliana whispered.

Margotta sat down at the edge of the bed. She briefly looked at the window where a small, black bird was making some noise beating its wings against the glass outside.

"What do you mean it's over?" Margotta said after the bird flew away.

"It's over for me, my darling."

"What are you saying?" Margotta raised her voice. "You're talking nonsense. Cheer up baby! You'll be just fine. You need to rest a few days, that's all... Look!" She pointed at the window. "You see? The storm is gone. The sun is shining again... It's a pretty afternoon."

"Yes, it is... for you... I hope... The medic was here shortly before you came... She checked my blood... Anyway, let's talk about you... Since I got seriously ill, I haven't been able to get out of bed... How many scheduled visits to the Lipoclinic have I missed? Three... Four... Really, I haven't counted... It's sure nice to see you again."

"You know, I don't move around too easily, either. Certainly not like even just a couple of years ago... Remember, how happy we used to be?"

Tliana sighed. "I remember, love... I was always very happy with you... You know, now that I've reached the end of my rope, I think I'll have to tell you something... Maybe it'll shock you... I think it's better to be honest about it." She turned her head away and looked at the wall.

"I'm listening," Margotta said after a period of silence. "Oh, baby, don't cry!" she added when she saw the teardrops on Tliana's face.

"I cheated on you." Tliana's voice turned shaky. "I admit... I cheated on you."

"You did?" Margotta's response was quiet.

"Not physically," Tliana continued. "Not physically... Although I don't think it makes a big difference... I cheated you in my mind... I cheated you in my dreams and in my fantasies... I just couldn't resign to the fact that life robbed me of real love... Whenever I was with you, I imagined you were a strong, athletic man."

Margotta sniffed a few times. "Guess what, baby!" she said when Tliana finished. "We're even... I did the same."

"Really..." Tliana looked at Margotta again. "I'm so glad to hear you say that... It really makes me feel more at peace."

"Come on, Tliana! We knew this all along. We just never talked about it. Obviously, it wouldn't have been realistic for me or for you to wait for Superman to come and take us away. No sane man would honestly want to make love to four hundred-pound fat-balls... It's just so unfortunate that we were born genetically ill."

"Unfortunate?" Tliana took over trying to sound as indignant as her remaining strength allowed. "I say it is injustice! We're victims, my love... We're victims... All that experimentation... All that gene manipulation... All those synthetic medications... Well, we're the results."

"Hey, dig this! I guess you haven't been listening to the news lately."

"Who cares?"

"You might want to hear this, baby... Talking about experimentation and gene manipulation... Guess who used these exact same words last night on the evening TV news!"

"No idea."

"The Chancellor himself... I couldn't believe my ears. He gave an interview to a reporter from an independent station. That itself was news already. He talked about what he thought was wrong with life on our planet. He said he just began to realize that things have gone terribly wrong during the last few generations and that something would have to be done to correct the problems."

"Well, at least there is some hope for you, my love," Tliana whispered gasping for air.

Margotta made a face and shook her head. "No such luck, baby! No such luck... The hope ended with the morning news."

"What... happened?"

"The Chancellor was found dead in the morning. He was shut in the head on his way to his office... It's very strange, to say the least. Not long ago, an armed young man eluded security and penetrated the building. He got as far as the Chancellor's office. During that incident the Chancellor was able to defend himself and killed the intruder. Just between you and me, at that time I felt sorry for the young man. However, today, I mourn the Chancellor... Tliana! Are you listening?"

Tliana's eyes were closed. She opened her mouth trying to say something but she remained silent.

"Tliana..." Margotta grabbed her hand that lay motionless on the blanket.

Slowly, Tliana closed her mouth.

"Talk to me baby!" Margotta begged. "Say something, please."

There came no response.

Margotta reached for Tliana's wrist and tried to find her pulse. After a while, she gave up and placed Tliana's arm back on the bed. Slowly, she leaned over and touched her girlfriend's face with her own as her eyes got filled with tears.